Excel

YEAR 7

Problem Solving Workbook

ESSENTIAL skills

Get the Results You Want!

Allyn Jones

Reprinted 2014, 2018, 2021, 2022

Reviewed in 2024 for the NSW Curriculum and Australian Curriculum Version 9.0 changes

Reprinted 2025

ISBN 978 1 74125 407 5

Pascal Press
PO Box 250
Glebe NSW 2037
www.pascalpress.com.au

Publisher: Vivienne Joannou
Project editor: Rosemary Peers
Edited by Rosemary Peers
Answers checked by Peter Little
Cover, page design and typesetting by DiZign Pty Ltd
Printed by Vivar Printing/Green Giant Press

Introduction

This book has been specifically written for the **Australian Curriculum**.

The Australian Curriculum (Mathematics) covers six areas:

- Number
- Algebra
- Measurement
- Space
- Statistics
- Probability.

Students are encouraged to develop proficiency with mathematical concepts, skills, procedures and processes, and to use them to demonstrate mastery in mathematics as they solve problems.

This book provides strategies for students when applying mathematical concepts in a variety of routine and non-routine problems of increasing difficulty.

Please go to page 9 for Contents

KEY SKILL

11 Greatest common factor (Highest common factor)

HINTS

- A factor is a number that divides evenly into another number, e.g. Show that 7 is a factor of 42.
  As 42 ÷ 7 = 6 with no remainder, then 7 is a factor.
- The greatest common factor of two or more numbers is found by listing their factors and identifying the largest that is common, e.g. Find the greatest common factor of 16 and 36.
  16: 1, 2, **4**, 8, 16
  36: 1, 2, 3, **4**, 6, 9, 12, 18, 36 ∴ The greatest common factor is 4.

Reminder!

- Read the question carefully to identify what needs to be found.
- Reread the question when you've finished your answer to make sure that the question has been answered and your solution makes sense.

Examples

Mr Blake wants to divide the girls' choir into smaller groups. There are 12 sopranos and 20 altos. Each group needs the same number of each type of voice as every other group. What is the greatest number of groups that he can form?

Solution

Sopranos can be divided into 1, 2, 3, **4**, 6, or 12 groups.
Altos can be divided into 1, 2, **4**, 5, 10 or 20 groups.
∴ Mr Blake can form 4 groups.
(There will be 3 sopranos and 5 altos in each group.)

FOCUS on ...

1. The question: Asks you to find the largest number of groups that can be formed.
2. The information: Gives you the number of sopranos and altos.
3. Your working: The 12 sopranos will be split into equal sized groups. They can be split into a number of groups which are the factors of 12: 1, 2, 3, 4, 6, 12.
   Now the 20 altos are to be split into equal sized groups, which are the factors of 20: 1, 2, 4, 5, 10, 20. The greatest common factor gives the largest number of groups: the greatest common factor is 4.
4. Your answer: Now write a final statement: The greatest number of groups is 4.

Kristin wants to make bouquets from 16 red flowers and 24 white flowers. She wants to use the same number of each colour flower in each of her bouquets. What is the greatest number of bouquets Kristin can make?

Solution

Red: can be divided into 1, 2, 4, **8**, 16.
White: can be divided into 1, 2, 3, 4, 6, **8**, 12, 24.
∴ Kristin can make 8 bouquets.
(There will be 2 red flowers and 3 white flowers in each bouquet.)

FOCUS on ...

1. The question: Asks you to find the largest number of bouquets that can be formed.
2. The information: Gives you the number of red flowers and white flowers.
3. Your working: The 16 red flowers will be split into equal-sized bouquets. They can be split into a number of bouquets, which are the factors of 16: 1, 2, 4, 8, 16.
   Now the 24 white flowers are to be split into equal sized bouquets, which are the factors of 24: 1, 2, 3, 4, 6, 8, 12, 24. The greatest common factor is the largest number of bouquets; the greatest common factor is 8.
4. Your answer: Now write a final statement: The greatest number of bouquets is 8.

32 *Excel* Year 7 Problem Solving Workbook

THIS BOOK

Key skill

- Each unit focuses on one key skill from each of four sub-strands of the curriculum, e.g. Greatest common factor in the Number sub-strand.

Hints

- Students are given hints to help them solve the problems in the unit.
- These hints may also remind students of rules covered earlier in the book.

Reminder

- Students are given a reminder to always follow certain steps when they begin and end each question.

Examples

- Two examples are provided for each unit with worked solutions that students can follow.

Focus on ...

- A step-by-step method is provided for each example.
- These step-by-step methods are then used to solve the questions in the **Now try these!** section.
- Explanations and tips are provided in the methods to help students in their understanding and to encourage them to develop logical processes when solving problems in this unit.

A step-by-step guide to problem solving

Step 1: Focus on the question

- Read the question at least twice to make sure that you understand what you are being asked to solve.

Step 2: Focus on the information

- Concentrate on the relevant information. In some questions you may be given unnecessary information which will not be useful.
- Decide what you are looking for.

Step 3: Focus on your working

- There are many problem-solving strategies that you can use, including:
  - looking for a simpler problem of the same type
  - using easier numbers
  - working systematically
  - using the guess-and-check method
  - eliminating possibilities
  - drawing a picture or a diagram
  - drawing up a table
  - looking for a pattern
  - working backwards.

Step 4: Focus on your answer

- Make sure you write the correct units (e.g. minutes, metres) and ask yourself these questions:
  - Does the answer make sense? Is it a realistic answer?
  - Have I answered the question I was asked?

Now try these!

- All the questions in this section have been carefully written so that the exact same steps in solving them replicate the steps in the examples.
- The CHALLENGE question is a more challenging question where students will need to use slightly different steps to solve it.

Now try these!

1. The leaders of a youth group wants to split the boys and the girls into groups. There are 48 boys and 36 girls. Each group will have the same number of boys and the same number of girls. What is the largest number of groups that can be formed?

2. Anna is a florist and has 24 tulips and 36 lilacs to create bouquets. What is the largest number of bouquets she can make without having any flowers left over if each bouquet has the same number of each flower?

3. Nathan is stocking bathrooms at a hotel. He has 30 rolls of toilet paper and 20 cakes of soap. He wants to use all of the items and wants each of the bathrooms to be stocked with the same combination of supplies and have none left over. What is the greatest number of bathrooms he can stock?

4. A school has 60 pens and 80 pencils to make stationery packs for students. What is the maximum number of packs that can be made, if each pack is identical and all pens and pencils have been used?

5. Karen had a birthday party. She baked 48 cookies and 72 slices of pizza. Each person at the party ate the same number of cookies and the same number of slices of pizza. There was nothing left over. What was the most number of people at the party?

6. CHALLENGE Jessie is decorating a hall for her 18th birthday party. She wants to make identical balloon arrangements with each having the same number of each colour. If she has 30 pink balloons, 18 white balloons and 12 red balloons, what is the greatest number of arrangements she can make?

Answers pages 130–131

Revision Tests

- Pairs of revision tests appear regularly throughout the book to test students' understanding.
- Each pair has a test at both an Average and Challenging level of difficulty.

REVISION TEST 1 Level of difficulty—Average

1. Jerome recorded the number of points scored against his team in five games: 78, 111, 93, 87, 106. What was the total scored against his team in the five games?

2. The junior school has 465 students. An excursion to the zoo is planned and students are to be transported by buses which carry a maximum of 45 students plus two teachers. What is the smallest number of buses required?

3. A heavy fall of rain overnight filled the school's water tank to its 5600-litre capacity. Over the next five days the following amount of water was used: 80 L, 120 L, 175 L, 105 L and 205 L. If it did not rain during the five days, how much water remains in the tank?

4. Laura buys four DVDs at $14 each and six CDs at $12 each. What is her change from $150?

5. Jana tutors students in mathematics. Over four weeks she earns $120, $145, $105 and $110. What is her average weekly income?

6. Six times a week Tarek cycles 15 km around his local neighbourhood. If his goal is to cycle a total of 1350 km how many weeks will it take for Tarek to reach his goal?

Answers page 129

28 *Excel* Year 7 Problem Solving Workbook

Quick answers

- Students can quickly mark their work by referring to the quick answers in **bold** type.

Worked solutions

- Each question has a worked solution so that students can check incorrect answers or alternative ways to achieve the answer.

Worked solutions

NUMBER

Key Skill 1 Addition (pages 10–11)

1. **42 minutes**
Time = 14 + 6 + 22
= 42
∴ Connor takes 42 minutes.

2. **54 994**
21 122
10 380
+ 23 492
54 994
∴ the total population is 54 994.

3. **63**
Sam's age = 2
Sam's father's age = 2 + 28
= 30
Sam's grandfather's age = 30 + 33
= 63
∴ Sam's grandfather is 63 years old.

4. **323 m**
Height of Eureka Tower = 297
Height of Sydney Tower = 297 + 12
= 297 + 3 + 9
= 309
Height of Q1 = 309 + 14
= 323
∴ Q1 is 323 metres high.

5. **274**
69
70
68
+ 67
274
∴ Ryan's total was 274.

6. **84**
6 + 8 + 10 + 12 + 14 + 16 + 18 = 84
∴ Zac has completed 84 push-ups.

Key Skill 2 Subtraction (pages 12–13)

1. **368 000**
Consider the number of thousands:
547
− 179
368
∴ the difference is 368 000.

2. **1890 m**
9670 − 3600 − 4180
9670 − 3600 = 6070; 6070 − 4180 = 1890
∴ the plane is at 1890 m above sea level.

3. **600**
9000 − 3800 = 5200; 5200 − 4600 = 600
∴ 600 books are still to be delivered.

4. **238 000 000**
312
− 74
238
∴ Indonesia's population was 238 000 000.

5. **1983 km**
2793 − 435 = 2358; 2358 − 375 = 1983
∴ Hannah still has to travel 1983 km.

6. **$998 999 000**
1 000 000 000 − 1 000 000 = 999 000 000; 999 000 000 − 1 000 = 998 999 000
∴ $998 999 000 remains.

Key Skill 3 Multiplication (pages 14–15)

1. **$552**
Total amount = 23 × 4 × 6
= 92 × 6
= 552
∴ the total paid is $552.

2. **$156**
Total cost = 52 × 3
= 156
∴ the total cost is $156.

3. **960**
Total players = 16 × 10 × 6
= 160 × 6
= 960
∴ there are 960 players.

4. **$159**
Cost = $5.30 × 5 × 6
= $26.50 × 6
= $159
∴ the cost is $159.

126 *Excel* Year 7 Problem Solving Workbook

Contents

NUMBER

KEY SKILL

1 Addition

HINTS

- Remember that there are words and phrases used in word problems that tell you to add,
  e.g. sum, plus, total, increase, more, raise, combine, altogether, in all, both.
- Be careful to add digits with the same place values,
  e.g. 40 + 7 = 47
- Sometimes it is helpful to change the order of numbers in the sum,
  e.g. 17 + 45 + 13 + 25 = 17 + 13 + 45 + 25
  = 30 + 70
  = 100
- Also it might be easier if you write the numbers under each other when adding,
  e.g. 13 + 183 + 9

$$\begin{array}{r} 13 \\ 183 \\ +\quad 9 \\ \hline 205 \end{array}$$

Reminder!

- Read the question carefully to identify what needs to be found.
- Reread the question when you've finished your answer to make sure that the question has been answered and your solution makes sense.

Examples

Kahrin spent 40 minutes at her dance class on Monday, 55 minutes on Wednesday and 45 minutes on Saturday. What was the total amount of time she spent at the classes over the three days?

Solution

Total time = 40 + 55 + 45
= 140

∴ she spends 140 minutes (or 2 hours 20 minutes).

FOCUS on ...

1. The question: Asks you to find the total time spent.
2. The information: Gives you various times for each day.
3. Your working: You need to find the total, which means the numbers are to be added:
   40 + 55 + 45 = 140
4. Your answer: Make sure you write the correct units (here either minutes, or hours and minutes): Kahrin spends a total of 140 minutes.

Joanna drove her sportscar 23 150 km in the first year she bought it, 17 550 km in the second year, 16 120 km in the third year and 11 320 km in the fourth year. What was the total number of kilometres she drove in the four years?

Solution

Total distance = 23 150 + 17 550 + 16 120 + 11 320
= 68 140

∴ Joanna drove a total of 68 140 km.

FOCUS on ...

1. The question: Asks you to find the total distance.
2. The information: Gives you various distances for each year.
3. Your working: You need to find the total, which means the numbers are to be added:

$$\begin{array}{r} 23\,150 \\ 17\,550 \\ 16\,120 \\ +\ 11\,320 \\ \hline 68\,140 \end{array}$$

4. Your answer: Make sure you write the correct units: Joanna drove 68 140 km.

Now try these!

1. To get to school, Connor has a 14-minute walk, a 6-minute wait and then a 22-minute bus ride. How long does the whole trip take?

2. The populations of three South Australian cities are listed: Whyalla 21 122; Victor Harbor 10 380 and Mount Gambier 23 492. What is the total population of the three cities?

3. Sam is 2 years old. Her father is 28 years older than her. Sam's grandfather is 33 years older than her father. How old is Sam's grandfather?

4. Melbourne's Eureka Tower is the third tallest building in Australia. It is 12 metres shorter than Sydney Tower, which is 14 metres shorter than the Gold Coast's Q1 building. If Eureka Tower is 297 metres tall, how high is the Q1 building?

5. Ryan played four rounds of golf in a tournament. He had rounds of 69, 70, 68 and 67. What was the total of his four rounds?

6. CHALLENGE Zac completes six push-ups on Day 1. On Day 2 he completes eight push-ups, on Day 3, 10 push-ups and so on, increasing the number of sit-ups by two each day. By the end of Day 7, how many push-ups has Zac completed in total?

Answers page 126

KEY SKILL

2 Subtraction

HINTS

- There are words and phrases used in word problems that tell you to subtract,
  e.g. difference, less, decrease, reduce, dropped, lost, how many more, fewer, how many are left, remains.
- Take care to subtract digits with the same place values,
  e.g. 540 − 36 = 504
- It might be easier to write the numbers under each other when you are subtracting,
  e.g. 285 − 73

$$\begin{array}{r} 285 \\ -\ \ 73 \\ \hline 212 \\ \hline \end{array}$$

Reminder!

- Read the question carefully to identify what needs to be found.
- Reread the question when you've finished your answer to make sure that the question has been answered and your solution makes sense.

Examples

The total surface area of Earth is 510 000 000 km^2. The area that is covered by water is about 361 000 000 km^2. What is the approximate area of Earth covered by land?

Solution

Land area in millions = 510 − 361
= 149

∴ the area of land is 149 000 000 km^2.

FOCUS on...

1. The question: Asks you to find the area covered by land.
2. The information: Gives you the total area for the earth and the area of the water.
3. Your working: You need to find the area covered by the land. This can been found by subtracting two very large numbers. As both areas are in the millions, just subtract the number of millions:
   510 − 361 = 149
4. Your answer: Remember to include the millions:
   The area is 149 million, or 149 000 000 km^2.

A pair of jeans was originally priced at $99. The store held a discount sale which dropped the price by $15. Sarah used her discount card which took another $3 off the price. What did she pay for the jeans?

Solution

Cost in dollars = 99 − 15 − 3
= 81

∴ Sarah paid $81.

FOCUS on...

1. The question: Asks you to find the cost of the jeans.
2. The information: Gives you what the jeans originally cost, the amount of store discount and the discount she got from her card.
3. Your working: You need to find the cost of the jeans. 'Dropped by $15' means a subtraction, then 'another $3 off' means another subtraction:
   99 − 15 − 3 = 81
4. Your answer: Make sure you include the $ sign:
   Sarah paid $81.

Now try these!

1. Bethany lives in Newcastle which has a population of 547 000 people, while her brother lives in Geelong which has a population of 179 000 people. What is the difference between the two populations?

2. A plane is flying at 9670 metres above sea level. It then commences its descent, and after 10 minutes has dropped 3600 metres. After a further 10 minutes it has dropped another 4180 metres. At what height is the plane above sea level?

3. Members of a local club have 9000 telephone books to deliver to households. On Saturday they distribute 3800 and on Sunday another 4600. How many telephone books have yet to be delivered?

4. In mid-2011, the population of USA was estimated as 312 000 000, which was 74 000 000 more than Indonesia. What was the population of Indonesia at that time?

5. At 8 am Hannah left Adelaide to drive 2793 km. By midday she had travelled 435 km. In the afternoon she drove another 375 km. How far has Hannah still to travel?

6. CHALLENGE The treasurer announced a one-billion-dollar fund to spend on education. One million dollars was used to build a school library and one thousand dollars was given for playground improvement. How much money remains?

Answers page 126

KEY SKILL

3 Multiplication

HINTS

- Remember that there are words and phrases used in word problems that tell you to multiply,
  e.g. product, times, of, by, twice.
- When you multiply whole numbers by 10 you just write a 0 on the end of the number in the question. Multiplying by 100 you write 00 and so on,
  e.g. $45 \times 10 = 450$
- Sometimes a difficult multiplication can be made easier by doubling one number and halving the other,
  e.g. $48 \times 18 = 96 \times 9$
  $= 864$
- You can change the order of the numbers to help make a multiplication easier,
  e.g. $27 \times 5 \times 20 = 27 \times 100$
  $= 2700$
- It might be easier to write the numbers under each other when you are multiplying,
  e.g. 2560×11

$$\begin{array}{r} 2560 \\ \times \quad 11 \\ \hline 28\,160 \\ \hline \end{array}$$

Reminder!

- Read the question carefully to identify what needs to be found.
- Reread the question when you've finished your answer to make sure that the question has been answered and your solution makes sense.

Examples

Every day for 12 days Ryan sent 20 text messages. If each of the messages cost 15 cents, what was the total cost?

Solution

Total cost $= 12 \times 20 \times 15$
$= 3600$

$\therefore$ the cost will be 3600 cents, or $36.

FOCUS on ...

1. The question: Asks you to find the total cost.
2. The information: Gives you the number of messages each day, the number of days and the cost of each message.
3. Your working: You have a choice whether to keep your calculation in cents and then change to dollars at the end, or to change the 15c to $0.15 at the start. Here, let's keep it in cents.
   You can multiply the three numbers together:
   $12 \times 20 \times 15 = 3600$
4. Your answer: Remember to write the units— here you change the 3600 cents to dollars: The cost is $36.

Liam is a window cleaner and charges $5 for each window he cleans. The owner of a large building contracts him to clean 8 rows of 24 windows. How much will Liam charge?

Solution

Total charge $= 8 \times 24 \times 5$
$= 960$

$\therefore$ Liam will charge $960.

FOCUS on ...

1. The question: Asks you to find the amount that Liam charges.
2. The information: Gives you the number of windows in each row, the number of rows and the charge for each window.
3. Your working: You multiply 8 and 24 to find the number of windows and then this answer is multiplied by 5 to get Liam's charge.
   You can multiply the three numbers together:
   $8 \times 24 \times 5 = 960$
4. Your answer: Write a final statement:
   Liam will charge $960.

Now try these!

1. The cost to stay at a backpacker's hostel is \$23 per person per night. What is the total amount paid by four people for six nights?

2. Buses are used to transport students from seven schools to a careers expo. If 52 students travel on each bus and admission to the expo is \$3 per student, what would be the total cost of entry for one busload of students?

3. At the state age netball competition there are six age divisions participating. Each team has 10 players and there are 16 teams in each age division. What is the total number of players competing?

4. Tia-Jane's mother travels to and from work by train five days a week. Her daily return fare is \$5.30. What would her fares cost over a six-week period?

5. A cinema complex has eight identical cinemas. Each cinema has 20 rows of 24 seats each. How many seats are in the entire complex?

6. CHALLENGE A soft-drink company has seven identical trucks that make deliveries to supermarkets. Each truck holds 20 cartons. Each carton contains 24 boxes, and each box holds 12 bottles each. If each truck delivers a truckload of soft drink each day, how many bottles will be delivered in a five-day week?

Answers pages 126–127

KEY SKILL

4 Division with remainders

HINTS

- Remember that there are words and phrases used in word problems that tell you to divide,
  e.g. quotient, divide evenly, goes into, share, split, out of.
- When a number is a factor of another then there is no remainder.
- The remainder is what is left over when one number is divided by another,
  e.g. What is the remainder when 53 is divided by 6?
  $53 \div 6 = 8$ and remainder 5 $\quad \therefore$ remainder 5.
- When you divide you need to be careful about place value. Sometimes a zero appears in the answer,
  e.g. $318 \div 3 = 106$
- It might be easier to rewrite a division question,
  e.g. $476 \div 4$ $\qquad \begin{array}{r} 119 \\ 4\overline{)476} \end{array}$

Reminder!

- Read the question carefully to identify what needs to be found.
- Reread the question when you've finished your answer to make sure that the question has been answered and your solution makes sense.

Examples

Some parents have volunteered to drive competitors to a regional swimming carnival. There are 45 competitors and each car cannot take more than four students. What is the smallest number of cars needed?

Solution

$$\text{Total number of cars} = 45 \div 4$$
$$= 11\frac{1}{4}$$

$\therefore$ 12 cars are needed.

FOCUS on...

1. The question: Asks you to find the smallest number of cars.
2. The information: Gives you the number of students who can fit into each car and the total number of students.
3. Your working: Remember, whatever answer you get has to be a whole number because you cannot have a fraction of a car. You divide the numbers:
   $45 \div 4 = 11\frac{1}{4}$, or 11 with a remainder of 1.
4. Your answer: As 11 cars will not carry everyone, 12 cars are needed.

Small bags containing six kiwifruit are sold in a fruit shop. Jemma needs 50 kiwifruit. How many bags must she buy?

Solution

$$\text{Number of bags} = 50 \div 6$$
$$= 8\frac{1}{3}$$

$\therefore$ Jemma needs to buy 9 bags.

FOCUS on...

1. The question: Asks you to find the number of bags.
2. The information: Gives you the number of kiwifruit needed and how many are in each bag.
3. Your working: You divide the numbers:
   $50 \div 6 = 8\frac{1}{3}$, or 8 with a remainder of 2
4. Your answer: As 8 bags will not have enough kiwifruit, Jemma needs to buy 9. The final statement is:
   Jemma needs to buy 9 bags.

Now try these!

1. Chewy chocolate bars are sold in packets of eight. Lauren wants to give a chocolate bar to each of her 29 students. How many bags will Lauren need to buy?

2. A tour group consists of 43 tourists. The whole group take a chairlift ride up to the top of a mountain. The maximum number for each chairlift is six persons. What is the smallest number of chairlifts used by the tour group?

3. 180 students are required to make a bus trip. If a bus holds a maximum of 50 passengers, how many buses were hired?

4. There were 53 golfers participating in a charity golf day. All players used golf buggies for transport around the course. If golf buggies take a maximum of four golfers, what was the minimum number of golf buggies needed?

5. The world's biggest lamington contains 6500 eggs. If the eggs were bought in cartons of 12, how many cartons needed to be purchased?

6. CHALLENGE Audio cable is for sale in two lengths. A package of 100 metres is priced at $50 and 30 metres for $20. To provide data cabling for a school, Steele needs to buy a total of 540 metres. What is the smallest amount he will pay for the cable?

Answers page 127

KEY SKILL

5 Addition and subtraction

HINTS

- Make sure you know the **Number Hints** which appeared on the previous pages.
- Remember, you can use grouping symbols to give an order to the way you use mathematical operations—find the value of the expression inside the grouping symbols first,
  e.g. Find the difference between 12 and the sum of 7 and 3:
  $12 - (7 + 3) = 12 - 10$
  $= 2$

Reminder!

- Read the question carefully to identify what needs to be found.
- Reread the question when you've finished your answer to make sure that the question has been answered and your solution makes sense.

Examples

In one day, Sophie sleeps for 8 hours, attends school for 6 hours, travels for 2 hours and plays sport for an hour. How much time remains for other activities?

Solution

Remaining time $= 24 - (8 + 6 + 2 + 1)$
$= 24 - 17$
$= 7$

∴ the time remaining is 7 hours.

FOCUS on ...

1. The question: Asks you to find the number of hours remaining in the day.
2. The information: Gives you the number of hours spent on four activities. You know there are 24 hours in a day.
3. Your working: You can use grouping symbols to subtract the total hours of sleeping, school, travelling and sport from the 24 hours in the day:
   $24 - (8 + 6 + 2 + 1) = 24 - 17$
   $= 7$
4. Your answer: Now write a final statement, making sure to write the units:
   7 hours remain.

On Wednesday, Selena bought a book which had 445 pages and that day she read 48 pages. The following day she read 86 pages and on Friday read 75 pages. How many pages did she have left to read?

Solution

Remaining pages $= 445 - (48 + 86 + 75)$
$= 445 - 209$
$= 236$

∴ Selena had 236 pages left to read.

FOCUS on ...

1. The question: Asks you to find the number of pages remaining.
2. The information: Gives you the number of pages read over three days and the total number of pages in the book.
3. Your working: You can use grouping symbols to subtract the total pages read from the number of pages in the book:
   $445 - (48 + 86 + 75)$
   $$\begin{array}{r} 48 \\ 86 \\ +\ 75 \\ \hline 209 \end{array} \qquad \begin{array}{r} 445 \\ -\ 209 \\ \hline 236 \end{array}$$
4. Your answer: Now write a final statement:
   236 pages are left to read.

Now try these!

1. Tyler bought 20 litres of paint for some rooms in his house. He used 3 litres to paint the kitchen, 5 litres to paint the rumpus room and 2 litres to paint the hallway. How much paint remains?

2. Andrew set out on a driving holiday. He left home on Friday and drove 380 km. On Saturday he drove 410 km and on Sunday 290 km. On Monday he drove back home and found that he had travelled a total of 1300 km in the four days. How far did he travel on Monday?

3. Chris is learning to play golf and he bought a bag of 40 golf balls. On his first day of golf he lost six balls. When he played the following week he lost another five balls, and in the third week he lost another eight balls. How many balls does Chris have remaining?

4. In 2009, 453 people were killed on New South Wales roads. The numbers in each category were 210 drivers, 102 passengers, 69 motor cyclists and 13 pedal bicyclists. The remaining deaths were those of pedestrians. How many pedestrians died?

5. For her birthday, Taylah was given a petrol card to the value of $250. In three weeks she used the card to buy $40, $65 and $70 worth of petrol. What amount remains on the petrol card?

6. CHALLENGE Jason wants to walk 30 km in a week. On the Sunday he walks 1 km, on Monday 2 km, on Tuesday 3 km. If he increases the distance walked by an extra km each day, how far short of his goal will he be?

Answers page 127

KEY SKILL

6 Addition and multiplication

HINTS

- Make sure you know the **Number Hints** which appeared on the previous pages.
- You can use the order of operations rule—BODMAS (Brackets Orders Division Multiplication Addition Subtraction), e.g. $11 \times 3 + 5 \times 3 = 33 + 15$
  $= 48$
- The 'O' in BODMAS stands for Orders which means 'powers of' or indices or exponents.

Reminder!

- Read the question carefully to identify what needs to be found.
- Reread the question when you've finished your answer to make sure that the question has been answered and your solution makes sense.

Examples

Allanah bought four new tyres at $116 each and had her car serviced for $180. Calculate the total cost.

Solution

Total cost $= 116 \times 4 + 180$

$= 644$

$\therefore$ Allanah paid $644.

FOCUS on ...

1. The question: Asks you to find the total cost of the tyres and the car service.
2. The information: Gives you the number of tyres, the cost of each tyre and the cost of a service.
3. Your working: The tyre cost is multiplied by 4 and the cost of the car service is added: $116 \times 4 + 180 = 644$
4. Your answer: Now write a final statement, making sure to write the units: Allanah paid $644.

A grocer sells oranges in 3-kg bags and potatoes in 5-kg bags. Kate buys five bags of oranges and four bags of potatoes. Find the total mass of her purchases.

Solution

Total mass $= 5 \times 3 + 4 \times 5$

$= 15 + 20$

$= 35$

$\therefore$ the total mass is 35 kg.

FOCUS on ...

1. The question: Asks you to find the total mass.
2. The information: Gives you the mass of the two types of bags and the number of each type.
3. Your working: This means 5 lots of 3 plus 4 lots of 5: $5 \times 3 + 4 \times 5 = 15 + 20 = 35$
4. Your answer: Now write a final statement, making sure to write the units: The total mass is 35 kg.

Now try these!

1. Hayden bought a stapler which cost $7 and three boxes of staples at $2 each. What was the total cost of the items?

2. Lucinda bought eight DVDs at $12 each and five Blu-ray discs at $18 each. What was the total amount she paid?

3. Two brothers work at different fast food restaurants. Lachlan is paid $12 per hour while his younger brother Sam is paid $11 per hour. If during one week, Lachlan works for 9 hours and Sam works for 7 hours, what is their total wage?

4. On a soccer team, there are seven 12-year-olds and four 13-year-olds. What is the total of their ages?

5. Each morning, Allan swims in a 50-metre pool. He swims eight laps on Monday, Wednesday and Saturday. On Tuesday and Thursday he swims 12 laps. He does not swim on Friday and Sunday. What is the total distance that Allan swims each week?

6. CHALLENGE In a golf tournament, the winner earned $120 000. Two players were equal second and were paid $72 000 each. Six players tied for fourth and received $28 000 each. The tenth placed player was paid $12 000. What was the total prizemoney for the first ten placegetters?

Answers pages 127–128

KEY SKILL

7 Addition and division

HINTS

- Make sure you know the **Number Hints** which appeared on the previous pages.
- You can use the order of operations rule—BODMAS (Brackets Orders Division Multiplication Addition Subtraction), e.g. $36 \div 3 + 15 \div 3 = 12 + 5$
  $= 17$
- The 'O' in BODMAS stands for Orders which means 'powers of' or indices or exponents.

Reminder!

- Read the question carefully to identify what needs to be found.
- Reread the question when you've finished your answer to make sure that the question has been answered and your solution makes sense.

Examples

Four friends went camping. Before they started, their backpacks had masses of 14 kg, 17 kg, 20 kg and 9 kg. If they shared the total mass evenly, how much would each carry?

Solution

Mass of each pack $= (14 + 17 + 20 + 9) \div 4$
$= 60 \div 4$
$= 15$

$\therefore$ the mass of each pack would be 15 kg.

FOCUS on ...

1. The question: Asks you to find the mass each person would carry.
2. The information: Gives you the masses of the four backpacks.
3. Your working: You have to find the average of the backpacks. The average is found by adding the four masses and then dividing this total by four:
   $(14 + 17 + 20 + 9) \div 4 = 60 \div 4 = 15$
4. Your answer: Now write a final statement, making sure to write the units:
   Each pack will have a mass of 15 kg.

Over five days Rachael cycled 12 km, 17 km, 5 km, 16 km and 10 km. What was the average distance she cycled each day?

Solution

Average $= (12 + 17 + 5 + 16 + 10) \div 5$
$= 60 \div 5$
$= 12$

$\therefore$ the average distance cycled was 12 km.

FOCUS on ...

1. The question: Asks you to find the average distance travelled per day.
2. The information: Gives you the distances travelled on the five days.
3. Your working: You have to find the average of the distances. The average is found by adding the five distances and then dividing this total by five:
   $(12 + 17 + 5 + 16 + 10) \div 5 = 60 \div 5 = 12$
4. Your answer: Now write a final statement, making sure to write the units:
   The average distance cycled was 12 km.

Now try these!

1. Three girls arrived at the airport for a flight. The airline told them that there was a baggage limit of 20 kg but the girls had suitcases of masses of 25 kg, 14 kg and 18 kg. They opened their bags and shared the total mass evenly. What was the new mass of each bag?

2. Our school has six Year 7 classes: 7W has 30 students; 7H 28 students; 7I 25 students; 7G 30 students; 7R 27 students and 7E 28 students. What is the average class size?

3. Two friends are going to the local show and decide to share their spending money. Barry has $76 and Bobby has $92. How much will each boy end up with?

4. A school is raising money for three villages in Tanzania. Each morning for a week students made a gold coin donation. The amounts collected were $86, $47, $39, $42 and $53. What amount will be given to each village?

5. In four games, Grace's netball team scored 23, 16, 31 and 18 goals. What was the average number of goals scored each game?

6. CHALLENGE At the end of their round, four golfers counted their golf balls. Janine had 13 balls, Yvette 16, Eric 10 and Darren had 18 balls. Eric had played so poorly he decided to give away all his balls so that the other three players would have an identical number of balls. How many balls did Janine receive?

Answers page 128

KEY SKILL

8 Multiplication and division

HINTS

- Make sure you know the **Number Hints** which appeared on the previous pages.
- You can use the order of operations rule—BODMAS (Brackets Orders Division Multiplication Addition Subtraction),
  e.g. $24 \div 6 \times 2 = 4 \times 2$
  $= 8$
  e.g. $24 \div (6 \times 2) = 24 \div 12$
  $= 2$
- The 'O' in BODMAS stands for Orders which means 'powers of' or indices or exponents.

Reminder!

- Read the question carefully to identify what needs to be found.
- Reread the question when you've finished your answer to make sure that the question has been answered and your solution makes sense.

Examples

Leon is on a course of antibiotics. His doctor has given him a bottle containing 48 tablets. He takes two tablets three times a day. How long will it take for Leon to complete the course of antibiotics?

Solution

Number of days $= 48 \div (2 \times 3)$
$= 48 \div 6$
$= 8$
$\therefore$ it will take 8 days.

FOCUS on ...

1. The question: Asks you to find the number of days until Leon finishes the bottle of tablets.
2. The information: Gives you the total number of tablets and how many he takes each day.
3. Your working: You have to find the number of tablets each day by multiplying and then the number of days is found by dividing. This can be found using grouping symbols:
   $48 \div (2 \times 3) = 48 \div 6$
   $= 8$
4. Your answer: Now write a final statement:
   It will take 8 days.

Amy needs $480 for a deposit for her overseas trip. Each week she works for 6 hours at $8 per hour. How many weeks will she have to work to earn the money for the deposit?

Solution

Number of weeks $= 480 \div (8 \times 6)$
$= 480 \div 48$
$= 10$
$\therefore$ Amy will have to work for 10 weeks.

FOCUS on ...

1. The question: Asks you to find the number of weeks Amy will have to work.
2. The information: Gives you the amount required, the hourly rate and the number of hours she works each week.
3. Your working: You have to find Amy's weekly pay by multiplying and then the number of weeks is found by dividing. This can be found using grouping symbols:
   $480 \div (8 \times 6) = 480 \div 48$
   $= 10$
4. Your answer: Now write a final statement:
   Amy will have to work for 10 weeks.

Now try these!

1 Alphonsus buys 120 bales of hay to feed his horses. He uses three bales of hay twice a day. How many days will the hay last?

2 Kenny runs 4 km three mornings every week. How many weeks will it take for him to cover a total of 60 km?

3 A naturopath gives Ray a 300-mL bottle of medicine to help with his blood pressure. He is instructed to take 10 mL twice a day. How many days will the bottle last?

4 Beck is paid $14 per hour and every week works a total of 6 hours. How many weeks will it take her to earn $840?

5 Pete has 1440 tomato seedlings. Every day he plants eight rows of 30 seedlings each. How many days will he need to plant all the seedlings?

6 CHALLENGE A truck is loaded with 20 cartons. Each carton holds 12 boxes. Each box holds 10 cans of beans. If 28 800 cans of beans are produced, how many trucks will be loaded?

Answers page 128

KEY SKILL

9 Various operations

HINTS

- Make sure you know the **Number Hints** which appeared on the previous pages.
- You can use the order of operations rule—BODMAS (Brackets Orders Division Multiplication Addition Subtraction).
- The 'O' in BODMAS stands for Orders which means 'powers of' or indices or exponents.

Reminder!

- Read the question carefully to identify what needs to be found.
- Reread the question when you've finished your answer to make sure that the question has been answered and your solution makes sense.

Examples

How much would Jason have left from $250 if he bought two shirts for $60 each, a belt for $35 and a pair of shoes for $55.

Solution

Total cost = 60 × 2 + 35 + 55
= 210
Change = 250 − 210
= 40
∴ the change is $40.

FOCUS on ...

1. The question: Asks you to find the change Jason receives.
2. The information: Gives you the cost of all the items Jason bought and the amount of money he gave to the sales assistant.
3. Your working: There are two steps to the solution. First, find the total cost of the items:
60 × 2 + 35 + 55 = 210
Now, to find the change, subtract this total from 250:
250 − 210 = 40
4. Your answer: Now write a final statement, making sure to write the dollar sign: The change is $40.

A shearer records the number of shorn sheep over five days: 210, 187, 202, 195 and 206. If he is paid $140 for every 100 sheep shorn, what is his total pay for the five days?

Solution

Total sheep = 210 + 187 + 202 + 195 + 206
= 1000
Payment = 1000 ÷ 100 × 140
= 1400
∴ the shearer is paid $1400.

FOCUS on ...

1. The question: Asks you to find the amount the shearer is paid.
2. The information: Gives you the number of sheep shorn on each of five days and the rate of pay for shearing.
3. Your working: There are two steps to the solution. First, find the total number of sheep shorn:
210 + 187 + 202 + 195 + 206 = 1000
Now, to find his pay, you need to find how many 100s are in 1000 and multiply this answer by 140:
1000 ÷ 100 × 140 = 1400
4. Your answer: Now write a final statement, making sure to write the dollar sign: The shearer is paid $1400.

Now try these!

1. How much change would Adam receive from $50 if he bought four mangoes at $3 each and two bags of potatoes at $6 each?

2. Ahmed is a bricklayer and is paid $550 for every 1000 bricks he lays. On Monday he laid 420 bricks, on Tuesday 380, on Wednesday 375, on Thursday 410 and on Friday 415. How much will he be paid for his five days' work?

3. Miriam spent $450 at a shopping centre. She bought three pairs of shoes at $85 a pair, two handbags at $45 each and a watch. How much did the watch cost?

4. Bill is a fencing contractor and is paid $90 for every 10 metres of completed fence. Over six days he completed the following lengths: 16 m, 32 m, 28 m, 19 m, 18 m and 27 m. What will be his payment for the six days?

5. Elias opens a 2-litre carton and pours 300 mL into each of two glasses and 250 mL into another glass. What amount remains in the carton?

6. CHALLENGE Last weekend Joshua visited his mother in the country. His car uses petrol at the rate of 9 litres for every 100 km he travels. On Saturday he travelled 440 km and on Sunday 360 km. If petrol cost $1.50 per litre, what did the weekend trip cost Joshua in petrol?

Answers page 129

REVISION TEST  Level of difficulty—Average

1. Jerome recorded the number of points scored against his team in five games: 78, 111, 93, 87, 106. What was the total scored against his team in the five games?

2. The junior school has 465 students. An excursion to the zoo is planned and students are to be transported by buses which carry a maximum of 45 students plus two teachers. What is the smallest number of buses required?

3. A heavy fall of rain overnight filled the school's water tank to its 5600-litre capacity. Over the next five days the following amount of water was used: 80 L, 120 L, 175 L, 105 L and 205 L. If it did not rain during the five days, how much water remains in the tank?

4. Laura buys four DVDs at $14 each and six CDs at $12 each. What is her change from $150?

5. Jana tutors students in mathematics. Over four weeks she earns $120, $145, $105 and $110. What is her average weekly income?

6. Six times a week Tarek cycles 15 km around his local neighbourhood. If his goal is to cycle a total of 1350 km how many weeks will it take for Tarek to reach his goal?

Answers page 129

REVISION TEST  Level of difficulty—Challenging

1. Lucas had some stickers. He gave 23 stickers to each of his four mates and still had 36 stickers left over. How many stickers did Lucas have to begin with?

2. 247 men and 353 women attended a concert. If they each paid $85 per ticket, what was the total amount paid?

3. Catherine goes shopping. A handbag costs three times as much as a dress. If the handbag costs $213, how much will she pay for two dresses and a handbag?

4. Thirty adults and some children attended a play. The price of admission for an adult was $12 and a child $8. If a total of $520 was collected for all tickets, how many children attended the play?

5. Sara bought 1250 grams of flour. She used some of the flour to make eight small cakes. If she had 290 grams of flour left, how much flour did she use for three of the small cakes?

6. For their birthday twins Jack and Jill were each given the same amount of money. Every day Jack spent $24 and Jill spent $20. When Jack had used up all of his money, Jill still had $36 remaining. How much money was originally given to each twin?

Answers pages 129–130

KEY SKILL

10 Lowest common multiple

HINTS

- When two numbers are multiplied together, the result is called a multiple.
- The multiples of a number are found by multiplying by the counting numbers,
  e.g. Write the first six multiples of 8.
  8, 16, 24, 32, 40, 48.
- The lowest common multiple of two or more numbers is found by listing their multiples and identifying the first that is common,
  e.g. Find the lowest common multiple of 3 and 4.
  3: 3, 6, 9, **12**, 15, …
  4: 4, 8, **12**, 16, …   ∴ the lowest common multiple is 12.

Reminder!

- Read the question carefully to identify what needs to be found.
- Reread the question when you've finished your answer to make sure that the question has been answered and your solution makes sense.

Examples

Jack, Jill and Jan share a bag of tokens equally and there are none left over. If Jeremy had joined the group and the four friends had shared the bag of tokens equally there would be no tokens left over. What is the smallest number of tokens possible in the bag?

Solution

If 3 children share: 3, 6, 9, **12**, 15 …
If 4 children share: 4, 8, **12**, 16, …
The smallest number that is common to both is 12.
∴ the smallest number of tokens is 12.

FOCUS on …

1. The question: Asks you to find the smallest number of tokens that is possible.
2. The information: Tells you the number of tokens is a multiple of 3 and a multiple of 4 because there is no remainder after division.
3. Your working: You need to consider the multiples of 3: 3, 6, 9, 12, 15, …
   The multiples of 4 are 4, 8, 12, 16, 20 …
   Look at both lists of multiples to find the lowest common multiple: the lowest common multiple is 12.
4. Your answer: Now write a final statement:
   The smallest number of tokens is 12.

Two taps are dripping. One tap drips every four seconds. The other tap drips every five seconds. If they drip at exactly the same instant, when is the next time they drip at the same instant?

Solution

First tap drips after 4, 8, 12, 16, **20**, … seconds.
Second tap drips after 5, 10, 15, **20**, … seconds.
The first number that is common is 20.
∴ the next time is at 20 seconds.

FOCUS on …

1. The question: Asks you to find the time it takes until the taps drip together.
2. The information: Tells you the first tap drips after multiples of four seconds and the second tap drips after multiples of five seconds.
3. Your working: You need to consider the multiples of 4: 4, 8, 12, 16, 20, …
   The multiples of 5 are 5, 10, 15, 20, 25, …
   Look at both lists of multiples to find the lowest common multiple: the lowest common multiple is 20.
4. Your answer: Now write a final statement:
   The next time they drip together is at 20 seconds.

Now try these!

1. Two boys share a bag of jelly beans equally and there are none left over. If another boy joins the group, the three boys share the bag of jelly beans equally and there are no jelly beans left over. What is the smallest possible number of jelly beans in the bag?

2. Two lights are blinking. One light blinks every three seconds. The other light blinks every five seconds. If they blink at exactly the same instant, when is the next time they blink at the same instant?

3. Bob and Ben are training for a triathlon. Bob rides every third morning and Ben rides every seventh morning. If they rode together this morning, how many more days until they ride again?

4. Three monkeys can share a pile of bananas equally. If another five monkeys arrive they can all share the pile of bananas equally. What is the smallest number of bananas in the pile?

5. Eight golfers buy a bag of golf balls. They divide the balls evenly and there are none left over. If two more golfers join them they share the bag of balls evenly among the 10 golfers and there are no balls left over. What is the smallest number of golf balls in the bag?

6. CHALLENGE Jack, Meg and Ryan phone their mother regularly. Jack phones every third night, Meg every fourth night and Ryan phones every sixth night. If they all ring their mother tonight, when is the next time they all ring on the same night?

Answers page 130

KEY SKILL

11 Greatest common factor (Highest common factor)

HINTS

- A factor is a number that divides evenly into another number, e.g. Show that 7 is a factor of 42.
  As 42 ÷ 7 = 6 with no remainder, then 7 is a factor.
- The greatest common factor of two or more numbers is found by listing their factors and identifying the largest that is common, e.g. Find the greatest common factor of 16 and 36.
  16: 1, 2, **4**, 8, 16
  36: 1, 2, 3, **4**, 6, 9, 12, 18, 36 ∴ The greatest common factor is 4.

Reminder!

- Read the question carefully to identify what needs to be found.
- Reread the question when you've finished your answer to make sure that the question has been answered and your solution makes sense.

Examples

Mr Blake wants to divide the girls' choir into smaller groups. There are 12 sopranos and 20 altos. Each group needs the same number of each type of voice as every other group. What is the greatest number of groups that he can form?

Solution

Sopranos can be divided into 1, 2, 3, **4**, 6, or 12 groups.
Altos can be divided into 1, 2, **4**, 5, 10 or 20 groups.
∴ Mr Blake can form 4 groups.
(There will be 3 sopranos and 5 altos in each group.)

FOCUS on ...

1. The question: Asks you to find the largest number of groups that can be formed.
2. The information: Gives you the number of sopranos and altos.
3. Your working: The 12 sopranos will be split into equal sized groups. They can be split into a number of groups which are the factors of 12: 1, 2, 3, 4, 6, 12.
   Now the 20 altos are to be split into equal sized groups, which are the factors of 20: 1, 2, 4, 5, 10, 20. The greatest common factor gives the largest number of groups: the greatest common factor is 4.
4. Your answer: Now write a final statement: The greatest number of groups is 4.

Kristin wants to make bouquets from 16 red flowers and 24 white flowers. She wants to use the same number of each colour flower in each of her bouquets. What is the greatest number of bouquets Kristin can make?

Solution

Red: can be divided into 1, 2, 4, **8**, 16.
White: can be divided into 1, 2, 3, 4, 6, **8**, 12, 24.
∴ Kristin can make 8 bouquets.
(There will be 2 red flowers and 3 white flowers in each bouquet.)

FOCUS on ...

1. The question: Asks you to find the largest number of bouquets that can be formed.
2. The information: Gives you the number of red flowers and white flowers.
3. Your working: The 16 red flowers will be split into equal-sized bouquets. They can be split into a number of bouquets, which are the factors of 16: 1, 2, 4, 8, 16.
   Now the 24 white flowers are to be split into equal sized bouquets, which are the factors of 24: 1, 2, 3, 4, 6, 8, 12, 24. The greatest common factor is the largest number of bouquets; the greatest common factor is 8.
4. Your answer: Now write a final statement: The greatest number of bouquets is 8.

Now try these!

1. The leaders of a youth group wants to split the boys and the girls into groups. There are 48 boys and 36 girls. Each group will have the same number of boys and the same number of girls. What is the largest number of groups that can be formed?

2. Anna is a florist and has 24 tulips and 36 lilacs to create bouquets. What is the largest number of bouquets she can make without having any flowers left over if each bouquet has the same number of each flower?

3. Nathan is stocking bathrooms at a hotel. He has 30 rolls of toilet paper and 20 cakes of soap. He wants to use all of the items and wants each of the bathrooms to be stocked with the same combination of supplies and have none left over. What is the greatest number of bathrooms he can stock?

4. A school has 60 pens and 80 pencils to make stationery packs for students. What is the maximum number of packs that can be made, if each pack is identical and all pens and pencils have been used?

5. Karen had a birthday party. She baked 48 cookies and 72 slices of pizza. Each person at the party ate the same number of cookies and the same number of slices of pizza. There was nothing left over. What was the most number of people at the party?

6. CHALLENGE Jessie is decorating a hall for her 18th birthday party. She wants to make identical balloon arrangements with each having the same number of each colour. If she has 30 pink balloons, 18 white balloons and 12 red balloons, what is the greatest number of arrangements she can make?

Answers pages 130–131

KEY SKILL

12 Unitary method A

HINTS

- The unitary method finds the value of one unit which is then used to get the answer. This 'unit' could be, for example, the cost of one item, e.g. If 3 apples cost 60 cents, what is the cost of 1 apple?

  Cost of 1 apple $= 60 \div 3$
  $= 20$

  $\therefore$ one apple costs 20 cents.

  Now, if the cost of 1 apple is 20 cents, find the cost of 4 apples.

  Cost of 4 apples $= 20 \times 4$
  $= 80$

  $\therefore$ four apples cost 80 cents.

Reminder!

- Read the question carefully to identify what needs to be found.
- Reread the question when you've finished your answer to make sure that the question has been answered and your solution makes sense.

Examples

Last Sunday, Bryce took his family to the footy. For lunch he bought five pies for $15. How much would he have paid for eight pies?

Solution

Cost of 5 pies = $15

Cost of 1 pie $= \$15 \div 5$
$= \$3$

Cost of 8 pies $= \$3 \times 8$
$= \$24$

$\therefore$ it would cost $24 for 8 pies.

FOCUS on ...

1. The question: Asks you to find the cost of eight pies.
2. The information: Gives you the cost of five pies.
3. Your working: Find the cost of 1 pie and then the cost of 5 pies.
   Write down what you know: 5 pies cost $15.
   Divide by 5 to get the cost of 1 pie (because 1 pie will cost less money than 5): 1 pie costs $3.
   To find the cost of 8 pies, multiply by 8: 8 pies will cost $24.
4. Your answer: Now write a final statement: The 8 pies will cost $24.

A group of pickers harvest 4 hectares of tomatoes in 20 hours. How long would it take to harvest 7 hectares of tomatoes?

Solution

Time for 4 hectares = 20 hours

Time for 1 hectare $= 20 \div 4$
$= 5$

Time for 7 hectares $= 7 \times 5$
$= 35$

$\therefore$ it will take 35 hours.

FOCUS on ...

1. The question: Asks you to find the time to harvest 7 hectares.
2. The information: Gives you the time taken to harvest 4 hectares.
3. Your working: Find the time to harvest 1 hectare, then the time taken for 7 hectares.
   Write down what you know: 4 hectares takes 20 hours.
   Divide 20 by 4 to find the time for 1 hectare (because 1 hectare will take less time than 4): 1 hectare takes 5 hours.
   To find the time for 7 hectares, multiply by 7: 7 hectares take 35 hours.
4. Your answer: Now write a final statement: It will take 35 hours.

Now try these!

1. Lina is organising a touch-football team. She buys eight T-shirts for $56. What would Lina pay for 10 T-shirts?

2. If four loaded trucks contain a total of 36 tonnes of sand, what would be the total mass of sand carried by six loaded trucks?

3. The hair on Symon's head grows 4 cm in eight weeks. At this rate how long would it grow in 10 weeks?

4. Nim bought 10 hamburgers which cost $37.50. What is the cost of six of the hamburgers?

5. A packet of three tennis balls sells for $7.20. What would be the cost of five similar tennis balls?

6. CHALLENGE Larry the lumberjack takes six minutes to saw a log into three pieces. How long will it take him to saw a similar log into six pieces?

Answers page 131

KEY SKILL

13 Unitary method B

HINTS

- Make sure you know the **Unitary method Hints** from Key Skill 12.
- The unitary method finds the value of one unit which is then used to get the answer. This 'unit' could be, for example, the time it takes for one person to do something,

  e.g. If 3 men take 4 hours to finish a job, how long will it take 1 person?

  Time for 1 person $= 4 \times 3$
  $= 12$

  $\therefore$ it takes one person 12 hours.

  Now, if a job takes 1 person 12 hours to complete, how long will it take 2 people?

  Time for 2 people $= 12 \div 2$
  $= 6$

  $\therefore$ it takes two people 6 hours.

Reminder!

- Read the question carefully to identify what needs to be found.
- Reread the question when you've finished your answer to make sure that the question has been answered and your solution makes sense.

Examples

Five boys can mow a paddock in 6 hours. How long will it take three boys to mow the same paddock?

Solution

Time for 5 boys $= 6$ hours
Time for 1 boy $= 6 \times 5$
$= 30$
Time for 3 boys $= 30 \div 3$
$= 10$

$\therefore$ it will take 10 hours.

FOCUS on ...

1. The question: Asks you to find the time for 3 boys to mow an area.
2. The information: Gives you the time taken for 5 boys to mow an area.
3. Your working: Find the time for 1 boy to mow the area, then the time for 3 boys.
   Write down what you know: 5 boys take 6 hours.
   Multiply 6 by 5 to find the time for 1 boy (because 1 boy will take more time): 1 boy takes 30 hours.
   To find the time for 3 boys, divide by 3: 3 boys take 10 hours.
4. Your answer: Now write a final statement:
   It will take 10 hours.

Six players hire a squash centre for a night. Each person pays $12 to cover the hiring. If there had been eight players, what would have been each player's share?

Solution

Share for 6 players $= \$12$
Share for 1 player $= \$12 \times 6$
$= \$72$
Share for 8 players $= \$72 \div 8$
$= \$9$

$\therefore$ it would cost $9 each.

FOCUS on ...

1. The question: Asks you to find the cost if 8 people shared the cost of squash court hire.
2. The information: Gives you the cost if 6 people shared the cost of squash court hire.
3. Your working: Find the cost for 1 person to hire the courts, then the cost for 8 people.
   Write down what you know: 6 players pay $12 each.
   Multiply $12 by 6 to find the cost of 1 player (because 1 player will have to pay more to hire the squash centre by him- or herself): 1 player pays $72.
   To find the cost for 8 players to hire the centre, divide by 8: 8 players would pay $9 each.
4. Your answer: Now write a final statement:
   It would cost $9 each.

Now try these!

1. If six men can load a truck in 30 minutes, how long would it take four men to load the same truck?

2. There is enough food to feed 12 soldiers for five days. How many days would the same amount of food last 15 soldiers?

3. A seniors group hires a coach for an outing and each person pays an equal amount towards the cost. If 40 seniors attend, the cost will be $4 each. What will be the cost for each person if only 32 seniors attend?

4. If 20 workers can complete a task in six days, how long would it take 24 workers working at the same rate?

5. The weekly rental is shared equally by the number of people living in a house. When there are four occupants each person pays $90 per week. What will be the share if there are six occupants?

6. CHALLENGE Michael filled a seed container and noticed that his eight cockatiels took five days to eat all the seed. He added two more cockatiels to the cage and filled the seed container again. How long would it take the cockatiels to eat all the seed?

Answers page 131

KEY SKILL

14 Unitary method C

HINTS

- Make sure you know the **Unitary method Hints** which appeared on previous pages.
- Setting out is important and you can list your solution (as in examples below) or draw a table to summarise your steps.

Reminder!

- Read the question carefully to identify what needs to be found.
- Reread the question when you've finished your answer to make sure that the question has been answered and your solution makes sense.

Examples

It takes six chickens two days to eat one bag of feed. How many days will it take four chickens to eat two bags of feed at the same rate?

Solution

Days for 6 chickens to eat 1 bag = 2

Days for 1 chicken to eat 1 bag = 2 × 6
= 12

Days for 1 chicken to eat 2 bags = 12 × 2
= 24

Days for 4 chickens to eat 2 bags = 24 ÷ 4
= 6

∴ it will take 6 days.

FOCUS on ...

1. The question: Asks you to find the time it takes 4 chickens to eat 2 bags of feed.
2. The information: Gives you the time it takes 6 chickens to eat 1 bag of feed.
3. Your working: Write down what you know: 6 chickens eat 1 bag in 2 days.
   Multiply 2 by 6 because 1 chicken will take more time: 1 chicken eats 1 bag in 12 days.
   Multiply 12 by 2 because 1 chicken will take more time to eat more bags: 1 chicken eats 2 bags in 24 days.
   Divide 24 by 4 because 4 chickens will take less time to eat the bags: 4 chicken eats 2 bags in 6 days.
4. Your answer: Now write a final statement:
   It will take 6 days.

If three furnaces use one tonne of coal in six hours, how long will it take two furnaces to use four tonnes?

Solution

Hours for 3 furnaces to use 1 t = 6

Hours for 1 furnace to use 1 t = 6 × 3
= 18

Hours for 2 furnaces to use 1 t = 18 ÷ 2
= 9

Hours for 2 furnaces to use 4 t = 9 × 4
= 36

∴ it will take 36 hours.

FOCUS on ...

1. The question: Asks you to find the time it takes for 2 furnaces to use 4 tonnes of coal.
2. The information: Gives you the time it takes 3 furnaces to use 1 tonne of coal.
3. Your working: Write down what you know: 3 furnaces use 1 tonne in 6 h.
   Multiply 6 by 3 because 1 furnace will take more time: 1 furnace uses 1 tonne in 18 h.
   Divide 18 by 2 because 2 furnaces will take less time: 2 furnaces use 1 tonne in 9 h.
   Multiply 9 by 4 because it will take more time to use more coal: 2 furnaces use 4 tonnes in 36 h.
4. Your answer: Now write a final statement:
   It will take 36 hours.

Now try these!

1. If three fire-fighters can clear 240 metres of a fire trail in two days, how many metres of fire trail can be cleared by four fire-fighters in three days, working at the same rate?

2. If 40 ants can eat one pie in two hours, how many ants can eat three pies in four hours?

3. If six workers complete 1 km of roadway in four days, how long would it take four workers to complete 3 km of roadway?

4. If three hens lay six eggs in four days, how many eggs would five hens lay in eight days?

5. If eight men mow four identical bowling greens in 20 minutes, how many men are needed to mow six similar sized bowling greens in one hour?

6. **CHALLENGE** 80% of the water in a bucket takes four hours to leak out through one small hole punched in the bottom of the bucket. A second hole of the same size is punched in the bottom of the bucket. Assuming water leaks out of the holes at a constant rate, how long would it take half of the remaining water to leak out?

Answers page 132

REVISION TEST  Level of difficulty—Average

1. In a cathedral's bell-tower there are three bells. One bell rings every 3 seconds, the second every 4 seconds and the third every 6 seconds. If they start ringing together, after how many seconds will they next ring together?

2. Jill is organising gift packs for students in Tanzania. She has 48 coloured pencils and 18 coloured pens. What is the largest number of gift packs she can make without having any pens or pencils left over if all gift packs are identical?

3. Bobby the barista can make 240 cups from 8 kg of coffee. How many cups could he make from 6 kg of coffee?

4. Klara is mowing a flat playing area. She mows one section which is 100 m^2 in 10 minutes. How long will it take her to mow a square of 360 m^2 if she works at the same rate?

5. Eight men can complete a job in six days. How long would it take three men to complete the same job?

6. If three cans of dog food will feed two dogs for one day, how many cans are required to feed eight dogs for a week?

Answers page 132

REVISION TEST  Level of difficulty—Challenging

1. A teacher asked a group of students to line up in twos. There was one student without a partner. The students then lined up in threes and there was one student left over. When the students lined up in fives there was still one person left over. What was the smallest number of students in the group?

2. If 24 people can complete a job in 10 days how many more people need to be employed to finish the job in four days?

3. Zoe has won some $10 phonecards as third prize in a competition she entered. She has decided to give them away to her friends. If she shares them among three friends there will be two left over. If she shares them among five friends there will be one left over. If she shares them among two friends there will be none left over. What is the smallest number of phonecards she could have won?

4. Owen harvested 2400 hectares of canola in 12 days using four machines. How long would it take to harvest 3600 hectares if he used six machines?

5. Brian is making up identical packs of muffins. He has baked eight chocolate, 16 strawberry, 12 banana muffins and some blueberry muffins. Each sample pack is to have a dozen muffins. What is the smallest number of blueberry muffins Brian could have baked?

6. Jake, Michael and Gavin walk up a flight of stairs. They all start on the first step. Jake goes up one step at a time. Michael goes up two steps at a time. Gavin goes up three steps at a time. Which is the fourth step they all step on together?

Answers page 133

KEY SKILL

15 Fractions: Addition

HINTS

- The top number in a fraction is called the numerator and the bottom number is the denominator.
- When adding fractions, if the denominators are the same, you just add the numerators,
  e.g. $\frac{3}{5} + \frac{4}{5} = \frac{7}{5} = 1\frac{2}{5}$
- When adding fractions, if the denominators are not the same, you first change the denominators to be the same,
  e.g. $\frac{2}{5} + \frac{3}{4} = \frac{8 + 15}{20} = \frac{23}{20} = 1\frac{3}{20}$

Reminder!

- Read the question carefully to identify what needs to be found.
- Reread the question when you've finished your answer to make sure that the question has been answered and your solution makes sense.

Examples

Elin jogs three-quarters of a kilometre on Thursday and two-thirds of a kilometre on Saturday. What is the total distance in the two days?

Solution

$$\text{Total distance} = \frac{3}{4} + \frac{2}{3} = \frac{9 + 8}{12} = \frac{17}{12} = 1\frac{5}{12}$$

$\therefore$ Elin has jogged $1\frac{5}{12}$ km.

FOCUS on ...

1. The question: Asks you to find total distance.
2. The information: Gives you the two distances that need to be added.
3. Your working: To add fractions with different denominators you need to find a common denominator. The smallest common multiple of 3 and 4 is 12. Make 12 the denominator for both fractions:
   $\frac{3}{4} = \frac{9}{12}$ [by mult. top and bottom by 3]
   $\frac{2}{3} = \frac{8}{12}$ [by mult. top and bottom by 4]
   As 9 plus 8 is 17, the total is $\frac{17}{12}$.
   Now change this improper fraction to a mixed numeral: 'how many 12s in 17?' The answer is 1 and 5 left over which is written as $1\frac{5}{12}$.
4. Your answer: Now write a final statement, remembering the units:
   Elin jogged $1\frac{5}{12}$ km.

At the supermarket Logan bought $4\frac{2}{3}$ kg of steak and $2\frac{4}{5}$ kg of sausages. What was the total mass of the meat?

Solution

$$\begin{aligned}\text{Total mass} &= 4\frac{2}{3} + 2\frac{4}{5} \\ &= 4 + 2 + \frac{2}{3} + \frac{4}{5} \\ &= 4 + 2 + \frac{10 + 12}{15} \\ &= 4 + 2 + \frac{22}{15} \\ &= 4 + 2 + 1\frac{7}{15} = 7\frac{7}{15}\end{aligned}$$

$\therefore$ Logan bought $7\frac{7}{15}$ kg of meat.

FOCUS on ...

1. The question: Asks you to find total mass.
2. The information: Gives you the two masses that need to be added.
3. Your working: To add mixed numerals you can add the whole numbers and then add the fractions. If the denominators are different, find a common denominator: the smallest common multiple of 3 and 5 is 15. Make 15 the denominator for both fractions:
   $\frac{2}{3} = \frac{10}{15}$ [by mult. top and bottom by 5]
   $\frac{4}{5} = \frac{12}{15}$ [by mult. top and bottom by 3]
   As 10 plus 12 is 22, the total is $\frac{22}{15}$.
   Now change this improper fraction to a mixed numeral: 'how many 15s in 22?' The answer is 1 and 7 left over which is written as $1\frac{7}{15}$.
   The whole numbers are 4, 2, 1 which add to 7.
4. Your answer: Now write a final statement, remembering the units:
   Logan bought $7\frac{7}{15}$ kg.

Now try these!

1. Over two nights Ella studied for her science test. On the first night she studied for three-quarters of an hour and on the second night for four-fifths of an hour. How long did she study in total as a mixed numeral?

2. A mechanic worked three-and-a-quarter hours on a car on Wednesday and two-and-a-half hours on Thursday. What was the total number of hours spent on the car?

3. Lane walked $2\frac{1}{5}$ kilometres on Saturday and cycled $5\frac{1}{2}$ kilometres on Sunday. What was the total distance?

4. Mrs Thompson needs one-and-a-half cups of milk to make a cake and three-and-two-thirds cups of milk to make custard. What is the total quantity of milk she needs?

5. A family ordered two pizzas. Tony ate a quarter of the ham and pineapple pizza and two-thirds of the supreme pizza. What was the total amount of pizza he ate?

6. CHALLENGE Phil is monitoring the bottles of water he drinks. On Monday he drinks $3\frac{1}{2}$ bottles, on Tuesday $4\frac{1}{3}$ bottles and on Wednesday $2\frac{1}{4}$ bottles. What was the total number of bottles consumed?

Answers pages 133–134

KEY SKILL

16 Fractions: Subtraction

HINTS

- When subtracting fractions, if the denominators are the same, you just subtract the numerators, e.g. $\frac{3}{5} - \frac{1}{5} = \frac{2}{5}$
- When subtracting fractions, if the denominators are not the same, you first change the denominators to be the same. You also subtract the whole numbers, then the fractions,
  e.g. $3\frac{4}{5} - 2\frac{1}{3} = 1 + \frac{12-5}{15} = 1\frac{7}{15}$
  e.g. $3\frac{1}{10} - 1\frac{4}{5} = 3 - 1 + \frac{1}{10} - \frac{8}{10} = 2 - \frac{7}{10} = 1\frac{3}{10}$

Reminder!

- Read the question carefully to identify what needs to be found.
- Reread the question when you've finished your answer to make sure that the question has been answered and your solution makes sense.

Examples

Javier is riding from his home to his grandmother's house. The entire trip takes five hours. If he has been riding for two-and-a-quarter hours, how long has he yet to travel?

Solution

$$\text{Time remaining} = 5 - 2\frac{1}{4}$$
$$= 3 - \frac{1}{4}$$
$$= 2\frac{3}{4}$$

$\therefore$ Javier has $2\frac{3}{4}$ hours still to travel.

FOCUS on...

1. The question: Asks you to find the time that remains to complete the journey.
2. The information: Gives you the total time and the time already used.
3. Your working: To subtract a mixed numeral from a whole number first subtract the two whole numbers: $5 - 2\frac{1}{4} = 3 - \frac{1}{4}$
   Now, as you know that $1 - \frac{1}{4}$ is $\frac{3}{4}$, so $3 - \frac{1}{4} = 2\frac{3}{4}$
4. Your answer: Now write a final statement: Javier has $2\frac{3}{4}$ hours still to travel.

Amanda bought $6\frac{3}{4}$ metres of flyscreen. After replacing the screens on three windows, she had $1\frac{1}{2}$ metres left over. How much did she use?

Solution

$$\text{Total mass} = 6\frac{3}{4} - 1\frac{1}{2}$$
$$= 5 + \frac{3}{4} - \frac{1}{2}$$
$$= 5 + \frac{3-2}{4}$$
$$= 5 + \frac{1}{4}$$
$$= 5\frac{1}{4}$$

$\therefore$ Amanda used $5\frac{1}{4}$ metres.

FOCUS on...

1. The question: Asks you to find the length remaining.
2. The information: Gives you the total length and the length already used.
3. Your working: To subtract a mixed numeral from another mixed numeral first subtract the two whole numbers and then subtract the fractions:
   $6\frac{3}{4} - 1\frac{1}{2} = 5 + \frac{3}{4} - \frac{1}{2}$
   Now for the fractions, you can use a common denominator of 4, because it is the smallest common multiple of 4 and 2:
   $\frac{3}{4} - \frac{1}{2} = \frac{3-2}{4} = \frac{1}{4}$
   Adding whole number and fraction you have
   $5 + \frac{1}{4} = 5\frac{1}{4}$
4. Your answer: Now write a final statement, remembering the units:
   Amanda used $5\frac{1}{4}$ metres.

Now try these!

1. Larry can swim 2 km while Barry can only swim $\frac{3}{4}$ km. How much further than Barry can Larry swim?

2. Noah bought a roll of wire of length 10 metres. If he used $4\frac{3}{4}$ metres to fix a fence, how much remained on the roll?

3. Terry has a cow and a goat that he milks each morning. The cow provided $9\frac{1}{2}$ litres. His goat gave $6\frac{1}{4}$ litres less than the cow. How much milk did Terry get from the goat?

4. One morning, a bakery used $12\frac{3}{4}$ bags of white flour and $5\frac{1}{2}$ bags of wholemeal flour to make loaves of bread. How much more white flour was used than wholemeal flour?

5. Owen and Aaron have just arrived at their caravan park for their holidays. Owen had driven for nine-and-three-quarter hours while it took Aaron six-and-half hours. How much longer did Owen take?

6. CHALLENGE Ken, Len and Jen race around a large park. Ken finishes first, taking $8\frac{1}{4}$ minutes. Jen takes $9\frac{3}{4}$ minutes and Len was third taking $10\frac{1}{2}$ minutes. After he finished, how long did Ken have to wait until Len finished?

Answers pages 134–135

KEY SKILL

17 Fractions: Multiplication A

HINTS

- Any whole number has a denominator of 1, such as $5 = \frac{5}{1}$.
- When multiplying you can cancel vertically and diagonally across the multiplication sign,

e.g. $\frac{3}{5} \times \frac{5}{6}$

$= \frac{{}^1\cancel{3}}{{}_1\cancel{5}} \times \frac{\cancel{5}^1}{\cancel{6}_2}$

$= \frac{1}{2}$

e.g. Find $\frac{2}{3}$ of 180.

$= \frac{2}{{}_1\cancel{3}} \times \frac{\cancel{180}^{60}}{1}$

$= 120$

Reminder!

- Read the question carefully to identify what needs to be found.
- Reread the question when you've finished your answer to make sure that the question has been answered and your solution makes sense.

Examples

Jenny is competing in a fun-run. The distance of the race is 15 km. If Jenny has run two-fifths of the race, what distance has she already run?

Solution

Distance $= \frac{2}{5} \times 15$

$= \frac{2}{{}_1\cancel{5}} \times \frac{\cancel{15}^3}{1}$

$= 6$

$\therefore$ Jenny has already run 6 km.

FOCUS on ...

1. The question: Asks you to find the distance already completed.
2. The information: Gives you the distance of the race and the fraction completed.
3. Your working: To multiply by a fraction you can use one of these three methods:
   - * $\frac{2}{5}$ of $15 = \frac{2}{{}_1\cancel{5}} \times \frac{\cancel{15}^3}{1} = 2 \times 3 = 6$; or
   - * $\frac{2}{5}$ of 15 = one-fifth of 15×2 $= 15 \div 5 \times 2 = 6$; or
   - * using the fraction keys on the calculator = 6
4. Your answer: Now write a final statement, remembering the units: Jenny has run 6 km.

Michael is given a grant of $400 to spend on insulation for his house. He has already spent three-eighths of the grant. How much money has Michael remaining?

Solution

Fraction remaining $= 1 - \frac{3}{8}$

$= \frac{5}{8}$

Amount remaining $= \frac{5}{8} \times 400$

$= \frac{5}{{}_1\cancel{8}} \times \frac{\cancel{400}^{50}}{1}$

$= 250$

$\therefore$ Michael has $250 remaining.

FOCUS on ...

1. The question: Asks you to find the amount of money that has not been spent.
2. The information: Gives you the amount of the grant and the fraction already spent.
3. Your working: First, find the fraction that has not been spent by subtracting from one whole:
   $1 - \frac{3}{8} = \frac{5}{8}$
   To multiply by a fraction you can use three methods:
   - * $\frac{5}{8}$ of $400 = \frac{5}{{}_1\cancel{8}} \times \frac{\cancel{400}^{50}}{1} = 5 \times 50 = 250$; or
   - * $\frac{5}{8}$ of 400 = one-eighth of 400×5 $= 400 \div 8 \times 5 = 250$; or
   - * using the fraction keys on the calculator = 250
4. Your answer: Now write a final statement, remembering the units: Michael has $250 remaining.

Now try these!

1. A jar contained 72 marbles and three-quarters of them were blue. How many blue marbles were in the jar?

2. Two-thirds of the students in a class are going on an excursion. If there are 27 students in the class, how many students are not going on the excursion?

3. Leonard buys 60 cans of soft drink. Two-fifths of the cans are cola flavoured. How many cola-flavoured cans of soft drink did he buy?

4. There are 144 students in Year Seven. Three-eighths of the students have a sister. How many students do not have a sister?

5. A youth group runs a car wash to support a number of charities. The total raised was \$240. If $\frac{5}{12}$ of the money raised goes to the Cancer Council, how much will be given to the other charities in total?

6. CHALLENGE A fitness club has 570 members. Two in every three members are female. What is the number of male members in the fitness club?

Answers page 135

KEY SKILL

18 Fractions: Multiplication B

HINTS

- Make sure you know the **Fractions Hints** which appeared on previous pages.
- When multiplying mixed numerals you rewrite them as improper fractions,

  e.g. $1\frac{3}{5} \times 2\frac{1}{2} = \frac{8}{5} \times \frac{5}{2}$

  $= \frac{^4\cancel{8}}{_1\cancel{5}} \times \frac{\cancel{5}^1}{\cancel{2}_1}$

  $= 4$

Reminder!

- Read the question carefully to identify what needs to be found.
- Reread the question when you've finished your answer to make sure that the question has been answered and your solution makes sense.

Examples

Ivan bought a coil of rope $5\frac{1}{4}$ metres in length to use on his yacht. He ended up only using two-thirds of the coil. What length did he use?

Solution

$$\text{Length} = \frac{2}{3} \times 5\frac{1}{4}$$
$$= \frac{2}{3} \times \frac{21}{4}$$
$$= \frac{^1\cancel{2}}{_1\cancel{3}} \times \frac{\cancel{21}^7}{\cancel{4}_2}$$
$$= 3\frac{1}{2}$$

$\therefore$ Ivan used $3\frac{1}{2}$ metres of rope.

FOCUS on…

1. The question: Asks you to find the length of rope used.
2. The information: Gives you the length of rope and the fraction of the rope that was used.
3. Your working: To multiply by a fraction you can use one of these two methods:

   * $\frac{2}{3}$ of $5\frac{1}{4} = \frac{^1\cancel{2}}{_1\cancel{3}} \times \frac{\cancel{21}^7}{\cancel{4}_2} = \frac{7}{2} = 3\frac{1}{2}$; or

   * using the fraction keys on calculator $= 3\frac{1}{2}$
4. Your answer: Now write a final statement, remembering the units:

   Ivan used $3\frac{1}{2}$ metres of rope.

A movie lasted two-and-a-quarter hours. Jelena arrived late and only managed to see three-quarters of the film. How much of the film did she miss, in hours?

Solution

$$\text{Fraction missed} = 1 - \frac{3}{4}$$
$$= \frac{1}{4}$$
$$\text{Amount missed} = \frac{1}{4} \times 2\frac{1}{4}$$
$$= \frac{1}{4} \times \frac{9}{4}$$
$$= \frac{9}{16}$$

$\therefore$ Jelena missed $\frac{9}{16}$ of an hour.

FOCUS on…

1. The question: Asks you to find the length of the movie that she did not see.
2. The information: Gives you the length of the movie and the fraction of the movie watched.
3. Your working: First, find the fraction that she missed by subtracting from one whole:

   $1 - \frac{3}{4} = \frac{1}{4}$

   To multiply by a fraction you can use two methods:

   * $\frac{1}{4}$ of $2\frac{1}{4} = \frac{1}{4} \times \frac{9}{4} = \frac{9}{16}$; or

   * using the fraction keys on calculator $= \frac{9}{16}$.
4. Your answer: Now write a final statement, remembering the units:

   Jelena missed $\frac{9}{16}$ of an hour.

Now try these!

1. A school held a lap-a-thon to raise money. Two-thirds of the funds raised went to the library. A quarter of the library funds were used to buy a new photocopier. What fraction of the money raised by the school was used for the photocopier?

2. A ball of string is $10\frac{1}{2}$ metres long. Keith used $\frac{4}{7}$ of the string to tie up some garden plants. How much of the string was not used?

3. Three-quarters of the cupcakes sold at a fete had strawberry icing. One-third of the cupcakes with strawberry icing had coconut sprinkles. What fraction of all the cupcakes had strawberry icing and coconut sprinkles?

4. Stefan took three-and-two-thirds hours to complete a marathon. The winner took three-fifths of Stefan's time. What was the time for the winner, in hours?

5. Two-thirds of the children in a group have a pet dog. Of those who have a dog, three-quarters have a cat as well. What fraction of the children in the group have a dog and a cat?

6. CHALLENGE A third of the students in a class have blonde hair. A quarter of the blonde-haired students do not have blue eyes. What fraction of the students have blonde hair and blue eyes?

Answers pages 135–136

KEY SKILL

19 Fractions: Division

HINTS

- The reciprocal of a fraction is found by swapping the numerator and denominator,

  e.g. What is the reciprocal of $\frac{3}{5}$?

  Reciprocal is $\frac{5}{3} = 1\frac{2}{3}$
- When dividing by a fraction you multiply by the reciprocal,

  e.g. $\frac{2}{5} \div \frac{3}{10} = \frac{2}{{}_1\cancel{5}} \times \frac{\cancel{10}^2}{3}$

  $= \frac{4}{3}$

  $= 1\frac{1}{3}$

Reminder!

- Read the question carefully to identify what needs to be found.
- Reread the question when you've finished your answer to make sure that the question has been answered and your solution makes sense.

Examples

Pedro inherited a small fortune from an uncle's estate. He gave two-thirds of his fortune to his four children. If the money was shared equally, what fraction of Pedro's inheritance will each child receive?

Solution

$$\text{Amount} = \frac{2}{3} \div 4$$
$$= \frac{{}^1\cancel{2}}{3} \times \frac{1}{\cancel{4}_2}$$
$$= \frac{1}{6}$$

$\therefore$ each child will receive $\frac{1}{6}$ of Pedro's inheritance.

FOCUS on ...

1. The question: Asks you to find the fraction of the inheritance for each child.
2. The information: Gives you the fraction of the inheritance going to the children and the number of children.
3. Your working: To divide by a fraction you can use one of two methods:
   - $\frac{2}{3} \div 4 = \frac{2}{3} \div \frac{4}{1} = \frac{{}^1\cancel{2}}{3} \times \frac{1}{\cancel{4}_2} = \frac{1}{6}$; or
   - using the fraction keys on the calculator $= \frac{1}{6}$
4. Your answer: Now write a final statement: Each child will receive $\frac{1}{6}$.

Helene is to cut a roll of ribbon into lengths of half a metre. If the original ribbon was $7\frac{1}{2}$ metres long, how many lengths can she cut?

Solution

$$\text{Number of lengths} = 7\frac{1}{2} \div \frac{1}{2}$$
$$= \frac{15}{{}_1\cancel{2}} \times \frac{\cancel{2}^1}{1}$$
$$= 15$$

$\therefore$ Helene is able to cut 15 lengths of ribbon.

FOCUS on ...

1. The question: Asks you to find the number of lengths possible.
2. The information: Gives you the length of the original roll and the lengths of each small section.
3. Your working: To divide by a fraction you can use one of three methods:
   - $7\frac{1}{2} \div \frac{1}{2} = \frac{15}{2} \div \frac{1}{2} = \frac{15}{{}_1\cancel{2}} \times \frac{\cancel{2}^1}{1} = 15$; or
   - there are 14 halves in 7 so that there are 15 halves in $7\frac{1}{2}$; or
   - using the fraction keys on the calculator = 15
4. Your answer: Now write a final statement: Helene can cut 15 lengths.

Now try these!

1. Grace swims the same distance each morning and over five days swam a total of $2\frac{1}{2}$ km. How far did she swim each morning?

2. Theodore bought $4\frac{1}{2}$ kg of dog food. He gave $\frac{1}{6}$ kg of food each day to his Labrador. How many days did the food last?

3. Youseff bought a 2-kg packet of washing powder. If he uses $\frac{1}{5}$ kg for each washing load, how many loads can he wash?

4. Chen uses three-eighths of a bottle of orange juice to fill a glass. How many glasses can be filled using 12 bottles?

5. Mum's recipe for cookies uses one-and-two-thirds cups of flour for every batch. How many batches of cookies can Jack make using 10 cups of flour?

6. CHALLENGE Every weekday Jose walks to work using the same route and then catches a bus home. He calculated that the route was $3\frac{1}{2}$ km and that he had walked a total of 70 km. For how many weeks did Jose walk to work?

Answers pages 136–137

KEY SKILL

20 Fractions: Addition and subtraction

HINTS

- Make sure you know the **Fractions Hints** which appeared on previous pages.
- You can use grouping symbols to give an order to the way you use mathematical operations. Remember, find the value of the expression inside grouping symbols first,

  e.g. Find the difference between 12 and the sum of $\frac{1}{2}$ and $\frac{1}{4}$.

  $12 - (\frac{1}{2} + \frac{1}{4}) = 12 - \frac{3}{4} = 11\frac{1}{4}$

Reminder!

- Read the question carefully to identify what needs to be found.
- Reread the question when you've finished your answer to make sure that the question has been answered and your solution makes sense.

Examples

Doug uses one-quarter of a tank of petrol on Monday and two-thirds of a tank on Tuesday. If he started with a full tank, what fraction of the tank remains?

Solution

$$\begin{aligned}\text{Remaining fraction} &= 1 - (\frac{1}{4} + \frac{2}{3}) \\ &= 1 - \frac{3 + 8}{12} \\ &= 1 - \frac{11}{12} \\ &= \frac{1}{12}\end{aligned}$$

∴ one-twelfth of the tank remains.

FOCUS on ...

1. The question: Asks you to find the fraction of the tank that remains.
2. The information: Gives you the amount of petrol used as fractions of the tank.
3. Your working: First, find the total of the two fractions by adding and then the answer is found by subtracting. This can be found using grouping symbols:

   $1 - (\frac{1}{4} + \frac{2}{3}) = \frac{1}{12}$
4. Your answer: Now write a final statement:

   One-twelfth of the tank remains.

Jessica purchases 3 metres of cable to connect speakers to her amplifier. She uses $\frac{3}{4}$ metre for one speaker and $\frac{2}{5}$ metre for the other. What length of cable is unused?

Solution

$$\begin{aligned}\text{Fraction used} &= 3 - (\frac{3}{4} + \frac{2}{5}) \\ &= 3 - \frac{15 + 8}{20} \\ &= 3 - \frac{23}{20} \\ &= 3 - 1\frac{3}{20} \\ &= 2 - \frac{3}{20} \\ &= 1\frac{17}{20}\end{aligned}$$

∴ $1\frac{17}{20}$ m is unused.

FOCUS on ...

1. The question: Asks you to find the length of the cable that has not been used.
2. The information: Gives you the length of the original cable and the lengths of the two sections used.
3. Your working: First find the total of the two fractions by adding and then the answer is found by subtracting. This can be found using grouping symbols:

   $3 - (\frac{3}{4} + \frac{2}{5}) = 1\frac{17}{20}$
4. Your answer: Now write a final statement:

   There is $1\frac{17}{20}$ metres remaining.

Now try these!

1. In the footy season so far, Jack has scored half the goals and Kramer has scored a fifth of the goals. If Craig kicked the rest of the goals, what fraction did he score?

2. In one season, a hockey team won three-fifths of its games and drew one-tenth. What fraction of the games did the team lose?

3. A survey found that one-quarter of all motor vehicle accidents were caused by drivers over 60 years old and three-fifths were caused by drivers under the age of 25. What fraction of all accidents were caused by drivers 25 to 60 years old?

4. Madeline bought 5 metres of material. She used $1\frac{1}{2}$ metres for a skirt and $2\frac{1}{4}$ metres for a dress. What amount of material remains?

5. In a packet of mixed lollies, one-third are jelly beans, one-quarter are caramels, one-sixth are buttons and the rest are musks. What fraction are musks?

6. CHALLENGE A family bought three pizzas for dinner. They ate three-quarters of one of the pizzas, two-thirds of another and five-eighths of the remaining pizza. What fraction of a pizza was not eaten?

Answers pages 137–138

KEY SKILL

21 Fractions: Addition, subtraction and multiplication

HINTS

- Make sure you know the **Fractions Hints** which appeared on previous pages.
- You can use the order of operations rule—BODMAS (Brackets Orders Division Multiplication Addition Subtraction), e.g. $12 \times \frac{2}{3} + 40 \div \frac{1}{4} = 8 + 160$
  $= 168$

Reminder!

- Read the question carefully to identify what needs to be found.
- Reread the question when you've finished your answer to make sure that the question has been answered and your solution makes sense.

Examples

Forty-five boys in Year Seven play one sport only. A fifth of the boys play tennis and two-thirds play soccer, and the remainder play basketball. How many play basketball?

Solution

Fraction play basketball $= 1 - (\frac{1}{5} + \frac{2}{3})$

$= \frac{2}{15}$

Number of basketballers $= \frac{2}{15} \times 45$

$= 6$

∴ there are 6 basketballers.

FOCUS on ...

1. The question: Asks you to find the number of boys who play basketball.
2. The information: Gives you the total number of boys and the fractions of the boys who play tennis and soccer.
3. Your working: First, you need to find the fraction who play basketball. You need to add the tennis and soccer fractions together, and then subtract this fraction from one whole:
   $1 - (\frac{1}{5} + \frac{2}{3}) = 1 - \frac{3 + 10}{15} = 1 - \frac{13}{15} = \frac{2}{15}$
   Now, multiply this fraction by the total number of boys:
   $\frac{2}{15} \times 45 = \frac{2}{{}_{1}\cancel{15}} \times \frac{\cancel{45}^{3}}{1} = 6$
4. Your answer: Now write a final statement:
   There are 6 basketballers.

36 000 people voted in a recent election. Two-thirds voted for Ms White and a quarter voted for Mr Brown. If the rest of the people voted for Mr Black how many votes did he receive?

Solution

Fraction voted for Black $= 1 - (\frac{2}{3} + \frac{1}{4})$

$= \frac{1}{12}$

Number of voters $= \frac{1}{12} \times 36\,000$

$= 3000$

∴ 3000 people voted for Mr Black.

FOCUS on ...

1. The question: Asks you to find the number of votes for Mr Black.
2. The information: Gives you the total number of voters and the fractions of voters who voted for White and Brown.
3. Your working: First, you need to find the fraction who voted for Mr Black. You need to add Ms White's and Mr Brown's fractions together, and then subtract this fraction from one whole:
   $1 - (\frac{2}{3} + \frac{1}{4}) = 1 - \frac{8 + 3}{12} = 1 - \frac{11}{12} = \frac{1}{12}$
   Now, multiply this fraction by the total number of voters:
   $\frac{1}{12} \times 36\,000 = \frac{1}{{}_{1}\cancel{12}} \times \frac{\cancel{36\,000}^{3000}}{1} = 3000$
4. Your answer: Now write a final statement:
   3000 voted for Mr Black.

Now try these!

1. A pole 6 metres in length is placed in a dam. One-sixth of the pole is in the ground and one-half of the pole is in the water. What length of the pole is above the water level?

2. In a school of 720 students, five-eighths of the students travel by bus, one-third travel by car and the remainder walk to school. How many students walk?

3. A bag contains 48 balls. One-third of the balls are red, one-quarter are blue and the rest are green. How many green balls are in the bag?

4. On her citrus farm Chloe has 900 trees. One-third of the trees are orange trees, two-fifths are lemon trees and the remainder are mandarine trees. How many mandarine trees has Chloe?

5. Hugo is paid $1200 a week. He pays one-quarter of his wage in tax and two-fifths is used to pay his rent. How much money does he have left over?

6. CHALLENGE A cruise ship has 3000 passengers. There are women, men and children on the ship. Three in every eight passengers are men and three in every five passengers are women. How many children are on the cruise?

Answers pages 138–139

KEY SKILL

22 Fractions using unitary method

HINTS

- The unitary method finds the value of one unit which is then used to get the answer. This 'unit' could be, for example, one-fifth of a quantity,
  e.g. If three-fifths of an object has a mass of 6 kg, what is the mass of one-fifth of the object?
  Mass of one-fifth of the object $= 6 \div 3$
  $= 2$
  $\therefore$ the mass of one-fifth of the object is 2 kg.
  Then the mass of four-fifths of the object is found by multiplying by 4.

Reminder!

- Read the question carefully to identify what needs to be found.
- Reread the question when you've finished your answer to make sure that the question has been answered and your solution makes sense.

Examples

Jacquie has a jar of beads. After using 48 beads to make a bracelet, she had a third of her beads remaining. How many beads were originally in the jar?

Solution

$$\text{Fraction used} = 1 - \frac{1}{3} = \frac{2}{3}$$

Beads in two-thirds of jar $= 48$
Beads in one-third of jar $= 48 \div 2$
$= 24$
Beads in three-thirds of jar $= 24 \times 3$
$= 72$
$\therefore$ there were 72 beads in the jar at the start.

FOCUS on ...

1. The question: Asks you to find the number of beads in the jar at the beginning.
2. The information: Gives you the number of beads used and the fraction of the total beads that remain in the jar.
3. Your working: The first thing you need to find is the fraction of beads that is equal to 48 beads:
   Subtract from one whole: $1 - \frac{1}{3} = \frac{2}{3}$
   Now, write down what is known: $\frac{2}{3}$ of jar $= 48$
   Dividing by 2: $\frac{1}{3}$ of jar $= 48 \div 2 = 24$
   Then, multiplying by 3 will give 3 thirds, which is 'a whole':
   3 thirds of jar $= 24 \times 3 = 72$
4. Your answer: Now write a final statement:
   There were originally 72 beads.

Bruno had a bucket of tomatoes. He used two-fifths of his tomatoes to make a relish. If he had 15 tomatoes left, how many did he use to make the relish?

Solution

$$\text{Fraction used} = 1 - \frac{2}{5} = \frac{3}{5}$$

Three-fifths of bucket $= 15$
One-fifth of bucket $= 15 \div 3$
$= 5$
Two-fifths of bucket $= 5 \times 2$
$= 10$
$\therefore$ Bruno used 10 tomatoes for the relish.

FOCUS on ...

1. The question: Asks you to find the number of tomatoes used to make the relish.
2. The information: Gives you the fraction of the tomatoes used and the number of tomatoes that remained.
3. Your working: The first thing you need to find is the fraction of the bucket that is equal to 15 tomatoes:
   Subtract from one whole: $1 - \frac{2}{5} = \frac{3}{5}$
   $\frac{3}{5}$ of bucket $= 15$
   Then, dividing by 3 will give one-fifth:
   $\frac{1}{5}$ of bucket $= 15 \div 3 = 5$
   Then, multiplying by 2 will give two-fifths, which is the fraction of the bucket Bruno used:
   $\frac{2}{5}$ of bucket $= 5 \times 2 = 10$
4. Your answer: Now write a final statement:
   Bruno used 10 tomatoes.

Now try these!

1. Nathan is reading a book. After reading three-eighths of the book he still has 250 pages remaining. What is the total number of pages in his book?

2. Yesterday, Wayne's water tank was three-fifths full. Overnight there was heavy rain which completely filled the tank. If 1800 L of water was added, how much water was in the tank before it rained?

3. Dorothy is travelling from Bundaberg to Rockhampton. She has travelled one-fifth of the distance but still has 200 km to travel. What is the total distance of her journey?

4. Four-fifths of the audience at a boxing match were male. If there were 120 females, how many males were at the match?

5. Sofia spent four-sevenths of her birthday money on a skateboard. If she had $90 remaining, how much money was she given for her birthday?

6. CHALLENGE Three-eighths of the students in a group cannot swim. There are eight more swimmers than non-swimmers. How many students are in the group?

Answers page 139

KEY SKILL

23 Ratio: Dividing quantities in a given ratio

HINTS

- A ratio compares quantities of the same unit, e.g. the ratio of 7 parts to 5 parts is written as 7:5
- Ratios are simplified in the same way that fractions are simplified—look for what is common and divide, e.g. simplify $12:$18 = 12:18 = 2:3
- The order is very important in ratio problems, e.g. the ratio of boys to girls is 3:4 which means that 3 parts are boys and 4 parts girls

Reminder!

- Read the question carefully to identify what needs to be found.
- Reread the question when you've finished your answer to make sure that the question has been answered and your solution makes sense.

Examples

A kiosk sold 120 cans in one day. The ratio of soft drinks to energy drinks was 2:3. How many cans of energy drink were sold?

Solution

$$\text{Total parts} = 2 + 3 = 5$$

$$\text{Energy drinks} = \frac{3}{{}_1\cancel{5}} \times \frac{\cancel{120}^{24}}{1} = 72$$

∴ 72 cans of energy drink were sold.

FOCUS on ...

1. The question: Asks you to find the number of energy drinks sold.
2. The information: Gives you the total number of cans sold, and the ratio of the two types of drinks.
3. Your working: Find the total number of parts. If the ratio is 2:3 then the total number of parts is 2 + 3 = 5.
   The order is important in ratio questions: energy drinks and 3 are mentioned second. This means 3 out of every 5 cans sold were energy drinks.
   You need to find $\frac{3}{5}$ of the total:
   $\frac{3}{5} \times 120 = \frac{3}{{}_1\cancel{5}} \times \frac{\cancel{120}^{24}}{1} = 72$
4. Your answer: Now write a final statement: The kiosk sold 72 cans of energy drink.

Bob and Bill start a business venture by investing $80 000 and $100 000 respectively. In the first year the profits amounted to $45 000. If they share the profits in the ratio of their investments, what is Bob's share of the profits?

Solution

$$\text{Ratio of investment} = \$80\,000:\$100\,000 = 4:5$$

∴ Bob:Bill is 4:5

$$\text{Total parts} = 4 + 5 = 9$$

$$\text{Bob's share} = \frac{4}{{}_1\cancel{9}} \times \frac{\cancel{45\,000}^{5000}}{1} = 20\,000$$

∴ Bob's share is $20 000.

FOCUS on ...

1. The question: Asks you to find Bob's share.
2. The information: Gives you the total profit and the amounts invested by the two men.
3. Your working: You have to find the ratio of their investments. You need to simplify:
   $80 000:$100 000 = 80 000:100 000 = 4:5
   Find the total number of parts. If the ratio is 4:5 then the total number of parts is 4 + 5 = 9.
   The order is important in ratio questions: Bob is mentioned first so that he has 4 shares out of 9.
   You need to find $\frac{4}{9}$ of the total:
   $\frac{4}{9} \times 45\,000 = \frac{4}{{}_1\cancel{9}} \times \frac{\cancel{45\,000}^{5000}}{1} = 20\,000$
4. Your answer: Now write a final statement, making sure to write the units: Bob receives $20 000.

Now try these!

1. On a school bus the ratio of male to female students is 5 : 3. If there is a total of 48 students on the bus, how many are females?

2. Kevin and Phillip made a total of 96 scones. For every two scones that Kevin made, Phillip made six. How many scones were made by Phillip?

3. A drink is made by mixing cordial and water in the ratio of 1 : 4. If Lee made 2.5 litres of drink, how much water is in the mixture?

4. Jason and Natasha bought a $10 lottery lottery ticket each week. Jason contributed $8 and Natasha the remainder. If they share a win of $25 000 in the ratio of their contributions, how much will Jason receive?

5. The ratio of the length to width of a rectangle is 6 : 5. If the perimeter is 66 cm, what is the width of the rectangle?

6. CHALLENGE For every six dozen white bread rolls prepared at a bakery in a particular week, three dozen multigrain rolls are made and half a dozen wholemeal rolls. If a total of 2280 bread rolls are baked, how many are not multigrain?

Answers page 140

REVISION TEST  Level of difficulty—Average

1. On Monday Rebecca jogged $4\frac{1}{5}$ km and on Wednesday $3\frac{3}{4}$ km. What was the total distance she jogged in the two days?

2. Tim purchased a $3\frac{1}{2}$ metre roll of wire. If he used three-quarters of the roll, what length remained?

3. Larry owns one-quarter of a business. Phillip owns one-fifth and Simone owns one-third. The bank owns the rest. What fraction of the business is owned by the bank?

4. Kim's petrol tank is one-quarter full. To fill the tank, Kim buys 36 litres. How many litres does the tank hold?

5. A drum is half full and contains 18 litres of oil. How much more oil is poured into the drum to make it two-thirds full?

6. The ratio of boys to girls at a party was 3 : 4. If there were 42 at the party how many were girls?

Answers pages 140–141

REVISION TEST  Level of difficulty—Challenging

1. Wang needs one-and-three-quarter cups of cream to make strawberry ice cream which serves seven people. How much cream is needed if he makes enough for 12 people?

2. On a trip to the United States, Taylor had $2500 spending money. She spent two-fifths of the money on clothes. One-quarter of the money she spent on clothes was used to buy four pairs of jeans of the same price. How much was the cost of a pair of jeans?

3. Trevor lost one-quarter of his marbles in a game. He gave three-fifths of the remaining marbles to his brother and put the rest of the marbles in four bags. If each bag contains 24 marbles, how many marbles did he have in the beginning?

4. Mark's father ate half of a block of chocolate. Mark ate half of what was left. Mark's mother ate half of what remained. The final four pieces were eaten by Mark's sister. How many pieces were in the original block?

5. Albert, Ben and Charles shared 72 stickers between themselves. Albert got three times as many stickers as Ben, while Ben got half as many as Charles. How many stickers did Ben receive?

6. Brodie grows lemons in his backyard. He sold one-third on the first day. He sold one-third of the remaining lemons on the second day. He still had 20 lemons remaining. How many lemons did he have originally for sale?

Answers pages 141–142

KEY SKILL

24 Decimals: Addition

HINTS

- There are words and phrases used in word problems that tell you to add,
  e.g. sum, plus, total, increase, more, raise, combine, altogether, in all, both.
- Take care to add digits with the same place values,
  e.g. 0.5 + 0.03 = 0.53
- Remember you can put zeros on the end of decimals to help with adding,
  e.g. 4.31 = 4.310 = 4.310000 = …
- It might be easier to write the decimals under each other when you are adding,
  e.g. 61.5 + 2.86

$$\begin{array}{r} 61.50 \\ +\ \ 2.86 \\ \hline 64.36 \end{array}$$

Reminder!

- Read the question carefully to identify what needs to be found.
- Reread the question when you've finished your answer to make sure that the question has been answered and your solution makes sense.

Examples

Amelia bought a new laptop for $1250, an external hard drive for $79.50 and a printer for $129.90. How much money did she spend altogether?

Solution

Total amount = 1250 + 79.5 + 129.9
= 1459.4

∴ Amelia spent $1459.40.

FOCUS on …

1. The question: Asks you to find the total amount spent.
2. The information: Gives you money spent on three items.
3. Your working: There are three prices that need to be added:

$$\begin{array}{r} 1250.0 \\ 79.5 \\ +\ 129.9 \\ \hline 1459.4 \end{array}$$

4. Your answer: Now write a final statement, making sure to write the units: Amelia spent $1459.40.

In the school's athletic carnival, Penny won the 100-metre race with a time of 12.65 seconds. Genevieve was second at 0.55 seconds behind Penny. What time did Genevieve run?

Solution

Time = 12.65 + 0.55
= 13.2

∴ Genevieve took 13.2 seconds.

FOCUS on …

1. The question: Asks you to find the time that Genevieve ran in her race.
2. The information: Gives the time that the first placegetter ran and the difference between the two times.
3. Your working: Genevieve took 0.55 seconds longer to run than Penny which means you need to add the time on to Penny's time:

$$\begin{array}{r} 12.65 \\ +\ \ 0.55 \\ \hline 13.20 \end{array}$$

4. Your answer: Now write a final statement, making sure to write the units: Genevieve took 13.2 seconds.

Now try these!

1. Serena has a mass of 46.8 kg and Patrick's mass is 51.3 kg. What is their total mass?

2. During a weekend outing Helena made three stops to buy petrol. At the first stop she bought 32.7 litres, at the second stop she bought 38.9 litres and at the third stop Helena bought 29.9 litres. How many litres of petrol did she buy altogether?

3. Kylie bought a pair of jeans for $69.95, a pair of shorts for $44.90 and a pair of sandals for $59.85. What was the total amount Kylie spent?

4. On her eleventh birthday Daniela's height was 1.31 metres. Over the following year, she grew by 0.08 metres, and in the next year by another 0.15 metres. How tall was Daniela on her thirteenth birthday?

5. A yacht sails three legs of a course in a race. The length of the legs are 4.8 km, 6.7 km and 4.05 km. What is the total distance the yacht sailed?

6. CHALLENGE Wen bought a box of muesli for $4.85, a bottle of orange juice for $2.50 and a carton of milk for $3.10. She received change of five cents. How much money did she hand over at the check-out?

Answers page 142

KEY SKILL

25 Decimals: Subtraction

HINTS

- There are words and phrases used in word problems that tell you to subtract,
  e.g. difference, less, decrease, reduce, dropped, lost, how many more, fewer, how many are left, remains.
- Take care to subtract digits with the same place values,
  e.g. 0.57 − 0.03 = 0.54
- Remember you can put zeros on the end of decimals to help with subtracting,
  e.g. 4.31 = 4.310 = 4.310 000 = …
- It might be easier to write the decimals under each other when you are subtracting,
  e.g. 61.5 − 2.86

$$\begin{array}{r} 61.50 \\ -\ \ 2.86 \\ \hline 58.64 \end{array}$$

Reminder!

- Read the question carefully to identify what needs to be found.
- Reread the question when you've finished your answer to make sure that the question has been answered and your solution makes sense.

Examples

Keith has a mass of 87.4 kg. He went on a diet and fitness program and lost 9.8 kg. What is his new mass?

Solution

New mass = 87.4 − 9.8
= 77.6

∴ Keith's new mass is 77.6 kg.

FOCUS on…

1. The question: Asks you to find Keith's new mass.
2. The information: Gives Keith's previous mass and the amount by which it has been reduced.
3. Your working: Keith has lost some weight so you need to subtract:

$$\begin{array}{r} 87.4 \\ -\ \ 9.8 \\ \hline 77.6 \end{array}$$

4. Your answer: Now write a final statement, making sure to write the units: Keith's new mass is 77.6 kg.

Leonie bought a roll of ribbon which was 4 metres in length. She cut 0.15 metres off the end. What length of ribbon remains on the roll?

Solution

Total amount = 4 − 0.15
= 3.85

∴ the length remaining is 3.85 metres.

FOCUS on…

1. The question: Asks you to find the length of the remaining ribbon.
2. The information: Gives the original length and the length that has been removed.
3. Your working: A section was cut off so you need to subtract:

$$\begin{array}{r} 4.00 \\ -\ \ 0.15 \\ \hline 3.85 \end{array}$$

4. Your answer: Now write a final statement, making sure to write the units: There is still 3.85 metres.

Now try these!

1. The reaction time of the athletes in a 100-metre race was recorded. Harold's reaction time was 0.32 seconds and Frederick's time was 0.21 seconds. What was the difference between their reaction times?

2. Benny has travelled 67.3 km of his 215 km journey. How far does he still have to travel?

3. In the morning the temperature was 11.7 degrees, but by 3 pm it had risen to 18.3 degrees. How much had the temperature risen?

4. Dianna has 2.6 litres of lemonade in a jug. If she pours 0.75 litres into a bottle, how much remains in the jug?

5. China and Japan are two of the world's biggest car manufacturers. In 2010, China produced 13.9 million cars while Japan produced 8.3 million. How many more cars did China produce than Japan?

6. CHALLENGE Brett had a block of wood 9 cm thick. From one side he planed off 0.3 cm and from the other side he planed off 0.45 cm. What was its final thickness?

Answers page 143

KEY SKILL

26 Decimals: Multiplication A

HINTS

- There are words and phrases used in word problems that tell you to multiply,
  e.g. product, times, of, by, twice.
- Decimal places are the number of digits that are to the right of the decimal point,
  e.g. 3.29 has two decimal places; 178.8054 has four decimal places.
- When you multiply decimals, the number of decimal places in the question will be identical to the number of decimal places in the answer,
  e.g. $3.2 \times 4 = 12.8$; $0.5 \times 0.03 = 0.015$
- The zero, or zeros, on the end of a decimal can be ignored,
  e.g. $5.4 \times 0.5 = 2.70 = 2.7$; $20 \times 0.005 = 0.100 = 0.1$
- You can ignore the decimal point and just multiply whole numbers first and then put the decimal point in at the end:
  e.g. 4.6×1.1. First find $46 \times 11 = 506$, then in goes the decimal point $\therefore$ 5.06
- It might be easier to write the decimals under each other when you are multiplying, e.g. 16.35×0.12

$$\begin{array}{r} 1635 \\ \times \quad 12 \\ \hline 19620 \end{array} \quad \therefore 1.962$$

Reminder!

- Read the question carefully to identify what needs to be found.
- Reread the question when you've finished your answer to make sure that the question has been answered and your solution makes sense.

Examples

Pens cost \$3.45 each. How much will George pay for six pens?

Solution

Cost $= 3.45 \times 6$
$= 20.7$
$\therefore$ The pens will cost \$20.70.

FOCUS on ...

1. The question: Asks you to find the cost of six pens.
2. The information: Gives the cost of a pen and the number of pens to be purchased.
3. Your working: You need to multiply the cost of a pen by 6:

$$\begin{array}{r} 345 \\ \times \quad 6 \\ \hline 2070 \end{array}$$

This means $345 \times 6 = 2070$.

In the question there were **2 digits** that followed decimal points; this means the answer will have **2 digits** that follow the decimal point:
$3.\underline{45} \times 6 = 20.\underline{70}$

4. Your answer: Now write a final statement, making sure to write the units: The pens will cost \$20.70.

A ream of paper has 500 sheets. If each sheet has a thickness of 0.121 mm, what is the thickness of the ream?

Solution

Ream thickness $= 0.121 \times 500$
$= 60.5$
$\therefore$ the ream is 60.5 mm thick.

FOCUS on ...

1. The question: Asks you to find the thickness of the ream.
2. The information: Gives the thickness of a sheet and the number of sheets.
3. Your working: You need to multiply the thickness of one sheet by 500:

$$\begin{array}{r} 121 \\ \times \quad 500 \\ \hline 60500 \end{array}$$

This means $121 \times 500 = 60500$.

In the question there were **3 digits** that followed decimal points; this means the answer will have **3 digits** that follow the decimal point:
$0.\underline{121} \times 500 = 60.\underline{500}$
Also, 60.500 can be rewritten as 60.5 because the 0s on the end are not important.

4. Your answer: Now write a final statement, making sure to write the units: The ream is 60.5 mm thick.

Now try these!

1. Every day a snail crawls 2.7 metres. How far would the snail crawl in five days?

2. Ewen buys 40 pavers to build a small path. If each brick paver has a mass of 2.15 kg, what is the total mass of the pavers?

3. Cassie buys six chocolate bars. If each bar costs $1.85 each, what is the total amount she will pay?

4. A fence-panel is 1.9 metres long. What is the total length of eight panels placed end to end?

5. One cubic centimetre of lead has a mass of 11.3 grams. Aaron is making sinkers to use when he goes fishing. If he makes a sinker that is two cubic centimetres, what is its mass?

6. CHALLENGE Mr Murray marked a maths test out of 80. To change the results to percentages, he multiplies each mark by 1.25. If Sandy scored 72 out of 80, what was her percentage?

Answers page 143

KEY SKILL

27 Decimals: Multiplication B

HINTS

- There are words and phrases used in word problems that tell you to multiply,
  e.g. product, times, of, by, twice.
- Decimal places are the number of digits that are to the right of the decimal point,
  e.g. 3.29 has two decimal places; 178.8054 has four decimal places.
- When you multiply decimals, the number of decimal places in the question will be identical to the number of decimal places in the answer,
  e.g. $3.2 \times 4 = 12.8$; $0.5 \times 0.03 = 0.015$
- The zero, or zeros, on the end of a decimal can be ignored,
  e.g. $5.4 \times 0.5 = 2.70 = 2.7$; $20 \times 0.005 = 0.100 = 0.1$
- You can ignore the decimal point and just multiply whole numbers first and then put the decimal point in at the end,
  e.g. 4.6×1.1. First find $46 \times 11 = 506$, then in goes the decimal point
  $\therefore$ 5.06
- It might be easier to write the decimals under each other when you are multiplying,
  e.g. 16.35×0.12

$$\begin{array}{r} 1635 \\ \times \quad 12 \\ \hline 19620 \end{array} \quad \therefore 1.962$$

Reminder!

- Read the question carefully to identify what needs to be found.
- Reread the question when you've finished your answer to make sure that the question has been answered and your solution makes sense.

Examples

Large king prawns are on sale for $24.90 per kg. If William buys 0.5 kg of the prawns, how much will he pay?

Solution

Cost $= 24.90 \times 0.5$
$= 12.45$

$\therefore$ the prawns will cost $12.45.

FOCUS on ...

1. The question: Asks you to find the cost of the prawns.
2. The information: Gives the cost of each kg of prawns and the amount of prawns.
3. Your working: You need to multiply the cost of the prawns for each kg by the number of kg.

$$\begin{array}{r} 2490 \\ \times \quad 5 \\ \hline 12450 \end{array}$$

This means $2490 \times 5 = 12450$.

In the question there were **3 digits** that followed decimal points; this means the answer will have **3 digits** that follow the decimal point:
$24.\underline{90} \times 0.\underline{5} = 12.\underline{450}$

Also, 12.450 can be written as 12.45 because the 0 on the end is not important.

4. Your answer: Now write a final statement, making sure to write the units: The cost will be $12.45.

Warren's hybrid car travels 25.2 km per litre. How far will it travel using 4.5 L?

Solution

Distance $= 25.2 \times 4.5$
$= 113.4$

$\therefore$ the car will travel 113.4 km.

FOCUS on ...

1. The question: Asks you to find the distance travelled.
2. The information: Gives the distance travelled on each litre and the number of litres.
3. Your working: You need to multiply the distance travelled on a litre by the number of litres.

$$\begin{array}{r} 252 \\ \times \quad 45 \\ \hline 1260 \\ 10080 \\ \hline 11340 \end{array}$$

In the question there were **2 digits** that followed decimal points; this means the answer will have **2 digits** that follow the decimal point:
We have $25.\underline{2} \times 4.\underline{5} = 113.\underline{40}$

Also, 113.40 can be written as 113.4 because the 0 on the end is not important.

4. Your answer: Now write a final statement, making sure to write the units: The car will travel 113.4 km.

Now try these!

1. Grant bought 0.8 kg of beans. If the beans were priced at $6.90 per kg, find the total cost.

2. Find the cost of 3.5 kg of chicken breasts at $8.90 per kg.

3. Camembert cheese costs $19.70 per kg. How much does Marianne pay for 0.4 kg of the cheese?

4. Ingrid is 1.2 times the height of her little brother Pedro. If Pedro's height is 134.5 cm what is the height of Ingrid?

5. An oil burner ran for 6.5 hours and used 1.4 litres per hour. What was the total amount of oil used?

6. CHALLENGE Neil is paid $14.50 per hour. How much is he paid if he works 8.5 hours?

Answers pages 143–144

KEY SKILL

28 Decimals: Division A

HINTS

- There are words and phrases used in word problems that tell you to divide,
  e.g. quotient, divide evenly, goes into, share, split, out of.
- Decimal places are the number of digits that are to the right of the decimal point,
  e.g. 3.29 has two decimal places; 178.8054 has four decimal places.
- When you divide a decimal by a whole number you have to take care with place value,
  e.g. $3.2 \div 4 = 0.8$; $63.6 \div 6 = 10.6$
- It might be easier to rewrite the decimal question,
  e.g. $25.84 \div 4$ $$\begin{array}{r} 6.46 \\ 4\overline{)25.84} \end{array}$$

Reminder!

- Read the question carefully to identify what needs to be found.
- Reread the question when you've finished your answer to make sure that the question has been answered and your solution makes sense.

Examples

Mitchell has 4.8 kg of coffee. He wants to put the coffee into six bags. How much will be in each bag?

Solution

$\text{Mass} = 4.8 \div 6$

$= 0.8$

$\therefore$ there is 0.8 kg in each bag.

FOCUS on ...

1. The question: Asks you to find the mass of each bag.
2. The information: Gives the total mass of coffee and the number of bags to be used.
3. Your working: You need to divide the total mass of coffee by the number of bags:
   $$\begin{array}{r} 0.8 \\ 6\overline{)4.8} \end{array}$$
4. Your answer: Now write a final statement, making sure to write the units: There will be 0.8 kg in each bag.

Dave, Tanya and Rob went to a restaurant for lunch. The total of the bill was $115.80. If they split the bill equally, how much did each person have to pay?

Solution

$\text{Cost} = 115.80 \div 3$

$= 38.60$

$\therefore$ each person paid $38.60.

FOCUS on ...

1. The question: Asks you to find the amount each person pays for lunch.
2. The information: Gives the total cost of lunch and the number of people who share the cost.
3. Your working: You need to divide the total cost by the number of people:
   $$\begin{array}{r} 38.60 \\ 3\overline{)115.80} \end{array}$$
4. Your answer: Now write a final statement, making sure to write the units: They paid $38.60 each.

Now try these!

1. Destiny ran a total of 23.1 km over three days. What was the average distance she ran each day?

2. Five friends went out to dinner. The cost of the meal was $428.50. If they shared the cost of the meal equally, how much did each person pay?

3. Aaron paid $58 for four adult tickets to the movies. What was the cost of each ticket?

4. Silas bought 11.25 kg of fertiliser to put in his three gardens. If he puts the same amount in each garden, how much fertiliser will each garden receive?

5. A biscuit factory used 34.8 kg of chocolate chips to make four batches of biscuits. How many kilograms of chocolate chips were used in each batch of biscuits?

6. CHALLENGE In New York Kath bought a multi-ticket to ride a subway for 20 days. If the ticket cost $76.20, how much did she pay for each day's travel?

Answers page 144

KEY SKILL

29 Decimals: Division B

HINTS

- There are words and phrases used in word problems that tell you to divide,
  e.g. quotient, divide evenly, goes into, share, split, out of.
- Decimal places are the number of digits that are to the right of the decimal point,
  e.g. 3.29 has two decimal places; 178.8054 has four decimal places.
- When you divide a decimal by a whole number you have to take care with place value,
  e.g. 3.2 ÷ 4 = 0.8; 63.6 ÷ 6 = 10.6
- It might be easier to rewrite the decimal question,
  e.g. 25.84 ÷ 4 $\qquad 4\overline{)25.84}$ = 6.46
- When you have to divide by another decimal you first need to change both numbers in the question so that you end up dividing by a whole number. You usually multiply by 10, 100, 1000, etc. so that the decimal point moves to the right a place, or two places or three places …
  e.g. 8.4 ÷ 0.4 = 84 ÷ 4 = 21; 6 ÷ 0.003 = 6000 ÷ 3 = 2000

Reminder!

- Read the question carefully to identify what needs to be found.
- Reread the question when you've finished your answer to make sure that the question has been answered and your solution makes sense.

Examples

An ice-cream scoop contains 0.08 litres of ice cream. How many scoops can Maria serve from a 4-litre container of ice-cream?

Solution

Scoops = 4 ÷ 0.08
= 400 ÷ 8
= 50

∴ Maria can serve 50 scoops.

FOCUS on …

1. The question: Asks you to find the number of scoops in the container.
2. The information: Gives the total amount of ice-cream and the amount in each scoop.
3. Your working: First, because you can only divide by a whole number you have to change the question. Multiply both numbers in the question by 100:
   4 × 100 is 400 and 0.08 × 100 is 8
   Now complete the solution:
   4 ÷ 0.08 = 400 ÷ 8 = 50
4. Your answer: Now write a final statement, making sure to write the units:
   Maria can serve 50 scoops.

A satellite takes 0.6 hours to orbit a small planet. How many complete orbits does the satellite make in 24 hours?

Solution

Number = 24 ÷ 0.6
= 240 ÷ 6
= 40

∴ the satellite will orbit 40 times.

FOCUS on …

1. The question: Asks you to find the number of orbits.
2. The information: Gives the total amount of time and the time taken for each orbit.
3. Your working: First, because you can only divide by a whole number you have to change the question. Multiply both numbers in the question by 10:
   24 × 10 is 240 and 0.6 × 10 is 6
   Now complete the solution:
   24 ÷ 0.6 = 240 ÷ 6 = 40
4. Your answer: Now write a final statement, making sure to write the units:
   The satellite will orbit 40 times.

Now try these!

1. Quentin has a piece of string which is 2.4 metres long. How many 0.8-metre lengths can be cut from the string?

2. Phillipa has a bucket containing 2 litres of milk. She pours the milk into identical glasses. If each glass contains exactly 0.25 litres, how many glasses has she filled?

3. When Harry walks, the length of his stride is 0.8 metres. How many strides must he take to travel 100 metres?

4. Darcy has a herd of dairy cows. A litre of Darcy's milk produces 0.04 kg of butter. How many litres of his milk are required to produce 1 kg of butter?

5. There are 1.6 km in a mile. How many miles are there in 80 km?

6. CHALLENGE Jim owns a cattle property and uses a 15-hectare paddock of pasture to fatten steers bought at auction. Each steer requires 0.75 hectare of pasture over three months during which they will be fattened. How many cattle will Jim put into the pasture?

Answers page 144

KEY SKILL

30 Decimals: Addition and subtraction

HINTS

- Make sure you know the **Decimals Hints** which appeared on previous pages.
- You can use grouping symbols to give an order to the way you use mathematical operations. Remember, find the value of the expression inside grouping symbols first,
  e.g. Find the difference between 12 and the sum of 7.1 and 3.9:
  $12 - (7.1 + 3.9)$
  $= 12 - 11$
  $= 1$

Reminder!

- Read the question carefully to identify what needs to be found.
- Reread the question when you've finished your answer to make sure that the question has been answered and your solution makes sense.

Examples

Toby bought a basketball for $29 and a pair of shorts for $19.90. How much change will he receive from a fifty-dollar note?

Solution

Change $= 50 - (29 + 19.90)$
$= 50 - 48.9$
$= 1.1$

$\therefore$ Toby will receive $1.10 change.

FOCUS on ...

1. The question: Asks you to find the amount of change.
2. The information: Gives the amount of money used to buy two items.
3. Your working: First, you find the total of the two decimals by adding and then the answer is found by subtracting using grouping symbols:
   $50 - (29 + 19.90)$

   $$\begin{array}{r} 29.00 \\ +\ 19.90 \\ \hline 48.90 \end{array} \qquad \begin{array}{r} 50.00 \\ -\ 48.90 \\ \hline 1.10 \end{array}$$

   This means, $50 - (29 + 19.90) = 50 - 48.90$
   $= 1.10$
4. Your answer: Now write a final statement, making sure to write the units: Toby will receive $1.10 in change.

Theodore's mass before he went on a diet and fitness program was 103.7 kg. In the first month he lost 4.8 kg, in the second month he lost 6.9 kg and in the third month he lost 8.1 kg. What was his mass at the end of the third month?

Solution

Mass $= 103.7 - (4.8 + 6.9 + 8.1)$
$= 103.7 - 19.8$
$= 83.9$

$\therefore$ his new mass was 83.9 kg.

FOCUS on ...

1. The question: Asks you to find Theodore's mass after he has been on a diet.
2. The information: Gives his original mass and the amount lost over three months.
3. Your working: First, you find the total of the three decimals by adding and then the answer is found by subtracting using grouping symbols:
   $103.7 - (4.8 + 6.9 + 8.1)$

   $$\begin{array}{r} 4.8 \\ 6.9 \\ +\ 8.1 \\ \hline 19.8 \end{array} \qquad \begin{array}{r} 103.7 \\ -\ 19.8 \\ \hline 83.9 \end{array}$$

   This means, $103.7 - (4.8 + 6.9 + 8.1)$
   $= 103.7 - 19.8$
   $= 83.9$
4. Your answer: Now write a final statement, making sure to write the units: Theodore's mass is 83.9 kg.

Now try these!

1. Pierre buys a maths study guide for $19.90 and a calculator for $18.60. How much change will he receive if he gives the cashier $40?

2. An electrician buys a 50-metre roll of electrical cable. In one week he uses 12.4 metres and then 18.7 metres in another week. What length of cable remains on the roll?

3. Zelda is given a $100 gift card for her birthday by her parents. She buys a bag for $38.50, a DVD for $19.90 and a magazine for $11.20. What amount of money remains on the card?

4. A school had a fundraising goal of $1000. The school raised $120.50 in the first week, $355.80 in the second week and $413.05 in the third week. How much more needs to be raised if the school is to meet its goal?

5. A blank DVD can store 4.3 gigabytes. If Toni transfers a 1.83 gigabyte folder of her photos and 1.75 gigabyte folder of her music to the DVD, what available space remains on the DVD?

6. CHALLENGE Inesa is paid $1040 for a week's work. She deducts $265.50 for tax, $142.60 for superannuation and $75.30 for her private health fund. How much will she have remaining?

Answers pages 144–145

KEY SKILL

31 Decimals: Addition, subtraction and multiplication

HINTS

- Make sure you know the **Decimals Hints** which appeared on previous pages.
- You can use the order of operations rule—BODMAS (Brackets Orders Division Multiplication Addition Subtraction), e.g. $12 \times 3.1 + 40 \div 8 = 37.2 + 5$
  $= 42.2$

Reminder!

- Read the question carefully to identify what needs to be found.
- Reread the question when you've finished your answer to make sure that the question has been answered and your solution makes sense.

Examples

Anne bought an 18-litre container of spring water for her family. Her son filled four 0.6-litre bottles and her daughter filled three 0.5-litre bottles. How much water is still in the container?

Solution

Used water $= 0.6 \times 4 + 0.5 \times 3$
$= 3.9$

Remainder $= 18 - 3.9$
$= 14.1$

∴ there is still 14.1 litres in the container.

FOCUS on...

1. The question: Asks you to find the amount of water remaining.
2. The information: Gives the original amount and the amount used by two children.
3. Your working: First, you work out the amount of water used, by multiplying and adding:
   $0.6 \times 4 + 0.5 \times 3$
   $$\begin{array}{r} 6 \\ \times\ 4 \\ \hline 24 \end{array} \qquad \begin{array}{r} 5 \\ \times\ 3 \\ \hline 15 \end{array}$$
   As 0.6×4 has 1 digit after a decimal point, then $0.6 \times 4 = 2.4$.
   Also, 0.5×3 has 1 digit after a decimal point, so $0.5 \times 3 = 1.5$.
   Now, add these two answers:
   $2.4 + 1.5$ $$\begin{array}{r} 2.4 \\ +\ 1.5 \\ \hline 3.9 \end{array}$$
   $2.4 + 1.5 = 3.9$
   Finally, subtract this total from 18:
   $18 - 3.9$ $$\begin{array}{r} 18.0 \\ -\ 3.9 \\ \hline 14.1 \end{array}$$
4. Your answer: Now write a final statement, making sure to write the units: There is still 14.1 litres.

Alfred bought three apples at 45 cents each and two bananas at \$1.20 each. How much change will he receive from \$10?

Solution

Cost $= 0.45 \times 3 + 1.2 \times 2$
$= 3.75$

Change $= 10 - 3.75$
$= 6.25$

∴ Alfred's change is \$6.25.

FOCUS on...

1. The question: Asks you to find the amount of change after buying some fruit.
2. The information: Gives the amount paid for some fruit and the cost of the fruit.
3. Your working: First, you need to find the total cost of the fruit by multiplying and adding:
   $0.45 \times 3 + 1.2 \times 2$
   $$\begin{array}{r} 45 \\ \times\ 3 \\ \hline 135 \end{array} \qquad \begin{array}{r} 12 \\ \times\ 2 \\ \hline 24 \end{array}$$
   As 0.45×3 has 2 digits after decimal points, then $0.45 \times 3 = 1.35$.
   Also, 1.2×2 has 1 digit after decimal points, so $1.2 \times 2 = 2.4$.
   Now, add these two answers:
   $1.35 + 2.4$ $$\begin{array}{r} 1.35 \\ +\ 2.40 \\ \hline 3.75 \end{array}$$
   $1.35 + 2.4 = 3.75$
   Finally, subtract this total from \$10:
   $10 - 3.75$ $$\begin{array}{r} 10.00 \\ -\ 3.75 \\ \hline 6.25 \end{array}$$
4. Your answer: Now write a final statement, making sure to write the units: Alfred's change is \$6.25.

Now try these!

1. On three days Jasmin walked 3.1 km each day and for four days she walked 2.4 km each day. Her goal for the week was 20 km. How far short of her goal was she?

2. A cricket club was given a $500 grant by a local RSL club. The club purchased 10 balls at $12.90 each and six bats at $49.90 each. What amount of the grant remains?

3. Cooper bought four tins of paint at $59.95 each, three paintbrushes at $11.90 each and a roller. If he spent a total of $294, how much did the roller cost?

4. For school Jamie bought four exercise books at $1.25 each, three pens at $2.05 each and a calculator. If she spent a total of $29.10, what was the cost of the calculator?

5. Four pies, two wraps and a carton of chocolate milk cost $23.65. If the cost of a pie was $3.75 and a wrap cost $2.90, what was the price of a carton of chocolate milk?

6. CHALLENGE A folder costs twice as much as a ruler which costs twice as much as a pencil. Sam bought twice as many rulers as pencils and twice as many folders as rulers. If a folder cost $1.20 and Sam bought two rulers, how much change did he get from $10?

Answers page 145

REVISION TEST  Level of difficulty—Average

1. It costs $3.20 to hire a taxi and then $1.80 for each kilometre travelled. How much will Shelby pay for a journey of 10 km?

2. Tom the butcher sells lamb mince for $9.90 per kilogram. What is the cost of 0.8 kg of the mince?

3. Vivienne takes 0.6 of an hour to ride around a circuit. If she maintains the same speed, how many laps will she complete in 2.4 hours?

4. A restaurant buys rice in 10-kg bags. If 3.2 kg was used on Monday and 2.9 kg on Tuesday night, how much rice remains?

5. Hennie bought three wraps at $3.20 each and two juices at $2.80 each. What is her change from $20?

6. Four overweight friends decide to attend a boot-camp to get fit. Craig has a mass of 102.5 kg, Jason 96.2 kg, Adam 98.6 kg and Brae 94.1 kg. What is the boys' average mass?

Answers pages 145–146

1. Tina wrote the numbers 4.8 and 0.6 on two cards. What is the difference between the quotient of the two numbers and their product?

2. Five friends entered the shot-put event at the school's athletics carnival and their average distance was 7.2 m. If four of the distances were 6.8 m, 7.5 m, 8.1 m and 7.2 m, what was the distance thrown by the fifth competitor?

3. A piece of wire is 29.4 metres long. Jack cuts the wire into equal lengths and has a length of wire left over which he throws away. The equal lengths are bent to form 24 squares with side length of 0.3 metres. What is the length of wire Jack threw away?

4. Susan fills her car with 28 litres of petrol which costs $44.80. How much will Noel spend if he fills his car with 50 litres?

5. The cost of edging a garden bed is $131.20. If the width of the garden bed is 8.4 metres what is its length if the edging costs $3.20 per metre?

6. In his will, a father leaves his three daughters 0.4 of his estate. Lacey receives 0.4 of the amount that Kacey receives and Tracey receives 0.4 of the amount that Lacey receives. If Tracey receives $16 000, what is the total value of the inheritance?

Answers page 146

KEY SKILL

32 Percentages: Finding one quantity as percentage of another

HINTS

- A percentage is a fraction with a denominator of 100, e.g. $\frac{48}{100}$ is written as 48%.
- One whole is 100%.
- To write a fraction as a percentage you multiply by 100%, e.g. What percentage is 43 out of 50?

$$\frac{43}{50} \times 100\% = \frac{43}{{}_{1}\cancel{50}} \times \frac{\cancel{100}^{2}}{1}\%$$
$$= 86\%$$

Reminder!

- Read the question carefully to identify what needs to be found.
- Reread the question when you've finished your answer to make sure that the question has been answered and your solution makes sense.

Examples

In a history test Jamie scored a mark of 28 out of 50. What is this result as a percentage?

Solution

$$\text{Result as percentage} = \frac{28}{50} \times 100\%$$
$$= 56\%$$

$\therefore$ Jamie scored 56% in the test.

FOCUS on ...

1. The question: Asks you to find Jamie's result as a percentage.
2. The information: Gives the mark Jamie earned and the mark that the test was marked out of.
3. Your working: First, write her result as a fraction the way the teacher would write on the top or front page of the test:
   28 out of 50 = $\frac{28}{50}$
   Then, to express this fraction as a percentage multiply by a whole (100%):
   $\frac{28}{50} \times 100\% = \frac{28}{{}_{1}\cancel{50}} \times \frac{\cancel{100}^{2}}{1}\% = 56\%$
4. Your answer: Now write a final statement, making sure to write the units: Jamie scored 56% in the test.

A car park has 60 parking spaces. If 27 of the spaces are occupied, what percentage of the car park is being used?

Solution

$$\text{Percentage used} = \frac{27}{60} \times 100\%$$
$$= 45\%$$

$\therefore$ 45% of the car park is used.

FOCUS on ...

1. The question: Asks you to find the percentage of spaces occupied.
2. The information: Gives the total number of car spaces and the number that are being used.
3. Your working: First, write the fraction of the spaces that are being used: 27 out of 60 = $\frac{27}{60}$.
   Then, to express this fraction as a percentage, multiply by a whole (100%):
   $\frac{27}{60} \times 100\% = \frac{{}^{9}\cancel{27}}{{}_{1}{}_{3}\cancel{60}} \times \frac{\cancel{100}^{5}}{1}\% = 45\%$
4. Your answer: Now write a final statement, making sure to write the units: 45% of the car park is used.

Now try these!

1. For Mother's Day, Herman gave his mother a box containing 20 chocolates. Sixteen of the chocolates were soft centres. What percentage of the box were soft centres?

2. Ninety students out of 120 students who attended the zone athletics carnival are now eligible to compete at the regional carnival. What percentage of the students can now compete at the regional carnival?

3. Last weekend Bruce went fishing. He cast out 60 times and caught three fish. What was his success rate as a percentage?

4. In a school survey we found that 240 out of 600 students walk to school. What percentage of the students surveyed walk to school?

5. Daniel counted 36 cockatoos in a tree. He 'scared' 27 of the birds and they flew away. What percentage of the cockatoos flew away?

6. **CHALLENGE** Three candidates contested a state government election and received 2229, 12 000 and 5771 votes respectively. What percentage of the total votes did the winning candidate get?

Answers pages 146–147

KEY SKILL

33 Percentages: Percentage change

HINTS

- Make sure you know the **Percentages Hints** from Key Skill 32.
- An increase is going up and a decrease is going down,
  e.g. If an amount changes from 60 to 85, what is the increase?
  Increase = 85 − 60 = 25
  e.g. If an amount changes from 100 to 65, what is the decrease?
  Decrease = 100 − 65 = 35
- Increases or decreases can be written as percentages,
  e.g. If an amount changes from 40 to 70 what is the percentage increase?
  Increase = 30
  $$\text{Percentage increase} = \frac{30}{40} \times 100\% = \frac{3}{\cancel{4}_1} \times \frac{\cancel{100}^{25}}{1}\% = 75\%$$
  ∴ 75%

Reminder!

- Read the question carefully to identify what needs to be found.
- Reread the question when you've finished your answer to make sure that the question has been answered and your solution makes sense.

Examples

Last year Pedro was paid $50 to look after his neighbours' dogs while they were away on a month's holiday. If this year he was paid $70, what was the percentage increase?

Solution

Increase = 70 − 50 = 20

$$\text{Percentage increase} = \frac{20}{50} \times 100\% = 40\%$$

∴ Pedro's pay was increased by 40%.

FOCUS on...

1. The question: Asks you to find the increase in Pedro's pay as a percentage.
2. The information: Gives Pedro's old pay and his new pay.
3. Your working: First, find the amount of increase by using subtraction:
   $70 − $50 = $20
   Next, write a fraction using the increase as the numerator and the original pay as the denominator: $\frac{20}{50}$
   Finally, to express this fraction as a percentage, multiply by a whole (100%):
   $$\frac{20}{50} \times 100\% = \frac{20}{\cancel{50}_1} \times \frac{\cancel{100}^2}{1}\% = 40\%$$
4. Your answer: Now write a final statement, making sure to write the units:
   His pay increased by 40%.

The price of a pair of netball shoes was reduced from $80 to $64. What is the decrease in price as a percentage?

Solution

Decrease = 80 − 64
= 16

$$\text{Percentage decrease} = \frac{16}{80} \times 100\% = 20\%$$

∴ the cost of the shoes dropped by 20%.

FOCUS on...

1. The question: Asks you to find the decrease in price as a percentage.
2. The information: Gives the old price and the new price.
3. Your working: First, use subtraction to find the amount of decrease:
   $80 − $64 = $16.
   Next, write a fraction using the decrease as the numerator and the original price as the denominator: $\frac{16}{80}$
   Finally, to express this fraction as a percentage multiply by a whole (100%):
   $$\frac{16}{80} \times 100\% = \frac{^2\cancel{16}}{_1\cancel{80}} \times \frac{\cancel{100}}{1}\% = 20\%$$
4. Your answer: Now write a final statement, making sure to write the units:
   The price of the shoes dropped 20%.

Now try these!

1. The price of a calculator increased from $20 to $24. What is the increase as a percentage of the original price?

2. A magazine drops in price from $10 to $9.00. Express this drop in price as a percentage of the original price.

3. Last month Angelo saved $70. If this month he saved $91, what is the increase in his monthly savings as a percentage?

4. A guitar was priced at $560. During a discount sale the price of the guitar dropped to $420. What was the percentage discount?

5. In the 2010 season the average home crowd at our stadium was 24 600. In the 2011 season the average increased to 36 900. What was the percentage increase?

6. CHALLENGE The population of a city increased from 175 000 to 210 000 in a decade. What was the average per cent increase **per year** throughout the decade?

Answers pages 147–148

KEY SKILL

34 Percentages: Fractions and percentages

HINTS

- Make sure you know the **Percentages Hints** which appeared on previous pages.
- To write a fraction as a percentage multiply by 100%,
  e.g. What percentage is 43 out of 50?

  $$\frac{43}{50} \times 100\% = \frac{43}{{}_{1}\cancel{50}} \times \frac{\cancel{100}^{2}}{1}\%$$
  $$= 86\%$$
- To write a percentage as a fraction divide by 100,
  e.g. What is 45% written as a fraction?

  $$45\% = \frac{45}{100}$$ (You need to simplify this fraction by cancelling.)
  $$= \frac{9}{20}$$

Reminder!

- Read the question carefully to identify what needs to be found.
- Reread the question when you've finished your answer to make sure that the question has been answered and your solution makes sense.

Examples

In 2003, the level of full immunisation of 12-month-old babies in Australia was 95%. What fraction of the babies were not immunised?

Solution

Percentage not immunised $= 100\% - 95\%$
$= 5\%$

Fraction not immunised $= \frac{5}{100}$
$= \frac{1}{20}$

$\therefore \frac{1}{20}$ of babies were not immunised.

FOCUS on ...

1. The question: Asks you to find the fraction of the babies that were not immunised.
2. The information: Gives the percentage of immunised babies.
3. Your working: First, subtract from a whole (100%) to find the percentage not immunised:
   $100\% - 95\% = 5\%$
   Now, to express this percentage as a fraction, write the percentage over 100 and cancel:
   $\frac{5}{100} = \frac{\cancel{5}^{1}}{\cancel{100}_{20}} = \frac{1}{20}$
4. Your answer: Now write a final statement, making sure to write the units:
   $\frac{1}{20}$ of babies were not immunised.

Two out of every five boys in a Year 7 class do not use their phones to listen to music. What percentage of the boys in the class use their phones to listen to music?

Solution

Fraction who do not listen to music $= \frac{2}{5}$

Fraction who listen to music $= 1 - \frac{2}{5}$
$= \frac{3}{5}$

Percentage who listen to music $= \frac{3}{5} \times 100\%$
$= 60\%$

$\therefore$ 60% of the boys listen to music on their phones.

FOCUS on ...

1. The question: Asks you to find the percentage of boys who use their phones to listen to music.
2. The information: Gives the fraction of boys not using their phone for music.
3. Your working: First, subtract from a whole to find the fraction who listen to music on their phones:
   $1 - \frac{2}{5} = \frac{3}{5}$
   Now, to express this fraction as a percentage, multiply by a whole (100%):
   $\frac{3}{5} \times 100\% = \frac{3}{{}_{1}\cancel{5}} \times \frac{\cancel{100}^{20}}{1}\% = 60\%$
4. Your answer: Now write a final statement, making sure to write the units: 60% of the boys listen to music on their phones.

Now try these!

1. Thirty per cent of the passengers on a bus were female. What fraction of the passengers on the bus were male?

2. A tub of ice-cream was opened and four-fifths of the ice-cream consumed. What percentage of the tub remained?

3. In 2010, 55% of the births in the local hospital were boys. What fraction of the babies were girls?

4. Rayyan attempted all 25 questions in a geography test. He got 7 of the questions incorrect. What percentage of the questions were answered correctly?

5. Three out of five pensioners reported that they had smoked when they were younger. What percentage of the pensioners had never smoked?

6. CHALLENGE A group of triathletes were interviewed to find which of the three legs was their favourite. 25% of the triathletes named the run leg while 35% said they favoured the bike section. What fraction of those interviewed named the swim leg as their favourite?

Answers page 148

KEY SKILL

35 Percentages: Using percentages A

HINTS

- Make sure you know the **Percentages Hints** which appeared on previous pages.
- To find a percentage of an amount, rewrite the percentage as a decimal or a fraction. In this book decimals will be used,
  e.g. Find 18% of 400.
  $0.18 \times 400 = 72.00$ $\therefore 72$
  e.g. Find 7% of 320.
  $0.07 \times 320 = 22.40$ $\therefore 22.4$

Reminder!

- Read the question carefully to identify what needs to be found.
- Reread the question when you've finished your answer to make sure that the question has been answered and your solution makes sense.

Examples

A tub of margarine has 4% total fat. How much fat is in a 450-gram tub?

Solution

Amount of fat = 4% of 450
= 0.04×450
= 18.00
= 18

$\therefore$ there is 18 grams of fat in the container.

FOCUS on ...

1. The question: Asks you to find the amount of fat in the tub of margarine.
2. The information: Gives the percentage of fat and the total mass of the margarine.
3. Your working: First, change the percentage to a decimal (or write it as a fraction): 4% is 0.04.
   You know that 'of' in maths means to multiply:
   0.04×450
   $$\begin{array}{r} 450 \\ \times \quad 4 \\ \hline 1800 \end{array}$$
   As there are two digits after the decimal point in the question, then there need to be two digits after the decimal point in the answer:
   $0.\underline{04} \times 450 = 18.\underline{00}$
4. Your answer: Now write a final statement, making sure to write the units: There is 18 grams of fat in the tub.

A school has 1100 students. Of these, 21% are in Year Seven. How many students are in Year Seven?

Solution

Number of students
= 21% of 1100
= 0.21×1100
= 231.00
= 231

$\therefore$ there are 231 students in Year Seven.

FOCUS on ...

1. The question: Asks you to find the number of students in Year Seven.
2. The information: Gives the number of students in the school and the percentage in Year Seven.
3. Your working: First, change the percentage to a decimal (or write it as a fraction):
   21% is 0.21
   You know that 'of' in maths means to multiply:
   0.21×1100
   $$\begin{array}{r} 21 \\ \times \ 1100 \\ \hline 23100 \end{array}$$
   As there are two digits after the decimal point in the question, then there need to be two digits after the decimal point in the answer:
   $0.\underline{21} \times 1100 = 231.\underline{00}$
4. Your answer: Now write a final statement, making sure to write the units: 231 students are in Year Seven.

Now try these!

1. A family spends 30% of its monthly income on home mortgage repayments. If the monthly income is \$7600, how much do they spend on their mortgage each month?

2. A holiday cruise was advertised for \$2400. The price is later discounted by 20%. By what amount was the price discounted?

3. Sixty per cent of the students in a Year Seven class have flown in an aeroplane. If there are 30 students in the class, what number have flown?

4. A case has 80 apples. When Owen checked the apples he found that 5% of the apples were rotten. How many rotten apples did he find in the case?

5. Bones make up about 15% of a person's body mass. If Wendy has a mass of 50 kg, find an approximation for the mass of her bones.

6. CHALLENGE To get her learner driver's licence, Serena sits a test consisting of 40 questions. She needs to score a result of at least 90% in the test. If all questions are of equal value, how many questions must she get correct?

Answers page 149

KEY SKILL

36 Percentages: Using percentages B

HINTS

- Make sure you know the **Percentages Hints** which appeared on previous pages.
- To increase a quantity by a percentage, first find the percentage and then add it on to the original amount,
  e.g. Increase 80 by 10%.
  Increase = 10% of 80
  = 0.10 × 80
  = 8
  New amount = 80 + 8
  = 88
- To decrease a quantity by a percentage, first find the percentage and then subtract it from the original amount.

Reminder!

- Read the question carefully to identify what needs to be found.
- Reread the question when you've finished your answer to make sure that the question has been answered and your solution makes sense.

Examples

The price of a bracelet was $64. If the price increased by 20%, what was the new price?

Solution

Increase = 20% of 64
= 0.2 × 64
= 12.8
∴ the price increased by $12.80.
New price = $64 + $12.80
= $76.80
∴ the new price was $76.80.

FOCUS on ...

1. The question: Asks you to find the new price after it has increased.
2. The information: Gives the old price and the increase as a percentage.
3. Your working: Change the percentage to a decimal (or write it as a fraction):
   20% is 0.2
   0.2 × 64

   ```
     64
   ×  2
    128
   ```

   As there is one digit after the decimal point in the question, then there needs to be one digit after the decimal point in the answer:
   0.2 × 64 = 12.8, which is $12.80
   You need to add the increase to the original price:
   $64.00 + $12.80

   ```
     64.00
   + 12.80
     76.80
   ```

4. Your answer: Now write a final statement, making sure to write the units: The new price was $76.80.

The population of a town in 2005 was 1550. By 2010, the population had dropped by 10%. What was the population in 2010?

Solution

Decrease = 10% of 1550
= 0.1 × 1550
= 155
∴ the population dropped by 155.
New population = 1550 − 155
= 1395
∴ the population was 1395 in 2010.

FOCUS on ...

1. The question: Asks you to find the new population after it has decreased.
2. The information: Gives the old population and the decrease as a percentage.
3. Your working: Change the percentage to a decimal (or write it as a fraction):
   10% is 0.1
   You know that 'of' in maths means to multiply:
   0.1 × 1550
   As there is one digit after the decimal point in the question, then there needs to be one digit after the decimal point in the answer:
   0.1 × 1550 = 155.0, which is 155.
   You need to subtract the decrease from the original population:
   1550 − 155

   ```
     1550
   −  155
     1395
   ```

4. Your answer: Now write a final statement, making sure to write the units: The population in 2010 was 1395.

Now try these!

1. In one week, Eisa's bean plant has increased in size by 30%. If last week it was 60 cm tall, what is its height this week?

2. Henry buys an orchard with 240 peach trees. If he reduces the number of trees by 20%, how many trees remain?

3. Michelle bought shares in a mining company that cost her $3.80 each. After a month each share had risen in price by 5%. What is the new value of each share?

4. Hala wanted to buy a diamond ring. The price was $480 but the jewellery store had a 20% discount sale on all its rings. What is the new price of the ring?

5. Grant bought a painting for $2100. After three years he sold the painting and made a profit of 40%. At what price did Grant sell the painting?

6. CHALLENGE A clothes store had a pair of jeans priced at $88. It increased the price by $22 but then discounted the price by 20%. What is the new price of the jeans?

Answers page 149

KEY SKILL

37 Percentages: Unitary method using percentages A

HINTS

- Make sure you know the **Percentages Hints** which appeared on previous pages.
- The unitary method finds the value of one unit which is then used to get the answer. This 'unit' could be 1% or 10%, or another percentage, e.g. 60% of a quantity is 42. What is the quantity?

  60% of quantity = 42

  10% of quantity = 42 ÷ 6

  = 7

  100% of quantity = 7 × 10

  = 70 ∴ the quantity is 70.

Reminder!

- Read the question carefully to identify what needs to be found.
- Reread the question when you've finished your answer to make sure that the question has been answered and your solution makes sense.

Examples

Andrew saves 30% of his income each week. If $360 is saved each week, what is Andrew's weekly income?

Solution

30% of income = $360

10% of income = $360 ÷ 3

= $120

100% of income = $120 × 10

= $1200

∴ Andrew is paid $1200 each week.

FOCUS on...

1. The question: Asks you to find Andrew's income.
2. The information: Gives the amount saved and the percentage of his income saved.
3. Your working: First, write down what you know:
   30% of income = $360
   Now, divide by 3 to get 10%:
   10% of income = $360 ÷ 3
   = $120
   To find his whole income you need 100%:
   100% of income = $120 × 10
   = $1200
4. Your answer: Now write a final statement, making sure to write the units: Andrew is paid $1200 each week.

Conan has been collecting stamps for a number of years. Eight per cent of the stamps in his collection are Australian. If Conan has 260 Australian stamps, how many stamps are in his collection?

Solution

8% of collection = 260

1% of collection = 260 ÷ 8

= 32.5

100% of collection = 32.5 × 100

= 3250

∴ Conan has 3250 stamps.

FOCUS on...

1. The question: Asks you to find the number of stamps in Conan's collection.
2. The information: Gives the number that are Australian and the percentage that are Australian.
3. Your working: First, write down what you know:
   8% of collection = 260
   Now, divide by 8 to get 1%:
   1% of collection = 260 ÷ 8 $\quad 8\overline{)260}$ = 32.5
   = 32.5
   To find the number in his collection you need 100%:
   100% of collection = 32.5 × 100
   = 3250
4. Your answer: Now write a final statement, making sure to write the units: Conan has 3250 stamps in his collection.

Now try these!

1. Barry forgot to close the farm gate and 8% of his herd of cows escaped. If 16 cows escaped, what was the total number of cows in Barry's herd before the escape?

2. Roberta buys a dress which has been discounted by 15%. If the price had dropped by $60, what was the original price of the dress?

3. Aaron has been given a pay rise of $60 per week. This is an increase in his pay of 5%. What was his weekly pay before the pay rise?

4. Since last year gold has increased in price by $120. If this is an increase of 20%, what was last year's price?

5. Kevin has 12% more stickers than his brother Paul. Paul has 48 stickers less than Kevin. How many stickers does Paul have?

6. CHALLENGE After a discount of 40% is allowed, the cost of insuring Yosef's house is $720. What would be the cost of insuring the house if he had no discount?

Answers page 150

KEY SKILL

38 Percentages: Unitary method using percentages B

HINTS

- Make sure you know the **Percentages Hints** which appeared on previous pages.
- If an amount (100%) increases by a percentage then you must add it on. If the amount instead decreases by a percentage you must subtract it,

  e.g. If an amount is increased by 30% what percentage of the original is the new amount?

  100% + 30% = 130%

  e.g. If an amount is decreased by 15% what percentage of the original is the new amount?

  100% − 15% = 85%

Reminder!

- Read the question carefully to identify what needs to be found.
- Reread the question when you've finished your answer to make sure that the question has been answered and your solution makes sense.

Examples

Alex is given a pay rise of 10%. His pay has increased to $880 per fortnight. What was his original pay?

Solution

New pay = 100% + 10%
= 110%

110% of old income = $880

10% of old income = $880 ÷ 11
= $80

100% of old income = $80 × 10
= $800

∴ Alex was originally paid $800.

FOCUS on...

1. The question: Asks you to find Alex's original pay.
2. The information: Gives the percentage rise and the new pay after the pay rise.
3. Your working: First, Alex increases his pay by 10%. Add 10% on to 100% to give 110%.
   This means that 110% of his old pay equals his new pay ($880):
   110% of old income = $880
   Now, divide by 11 to get 10%:
   10% of old income = $880 ÷ 11
   = $80
   To find his old income you need 100%:
   100% of old income = $80 × 10
   = $800
4. Your answer: Now write a final statement, making sure to write the units: Alex was originally paid $800.

A laptop drops in price by 20% and is now on sale for $1200. What was the original price of the laptop?

Solution

New price = 100% − 20%
= 80%

80% of old price = $1200

10% of old price = $1200 ÷ 8
= $150

100% of old price = $150 × 10
= $1500

∴ the original price was $1500.

FOCUS on...

1. The question: Asks you to find the original price of the laptop.
2. The information: Gives the percentage decrease and the new price after the drop in price.
3. Your working: First, the price drops by 20%. Subtract 20% from 100% to give 80%.
   This means that 80% of the old price equals the new price ($1200):
   80% of old price = $1200
   Now, divide by 8 to get 10%:
   10% of old price = $1200 ÷ 8
   = $150
   To find the old price you need 100%:
   100% of old price = $150 × 10
   = $1500
4. Your answer: Now write a final statement, making sure to write the units:
   The laptop was originally priced at $1500.

Now try these!

1. Four years ago William bought a painting. When he sold it last month he made a profit of 20%. If he sold it for $8400, what was William's original purchase price?

2. Malcolm inflated balloons for his party. After an hour 30% of the balloons had burst. If 140 balloons remained, how many were originally inflated?

3. Idris sells fruit and has some boxes of apples. She sells 40% of the apples and still has 360 apples remaining. How many apples did she have originally?

4. Sam had a truckload of soil delivered to his property. He used 20% of the load of soil to topdress his front yard. The remaining four tonnes of soil was used for gardens around his house. What was the mass of soil originally delivered?

5. Alma enjoyed her meal and gave a 10% tip to the waiter. She paid a total of $44, which included the tip. What was the cost of the meal before the tip?

6. CHALLENGE A car's fuel tank is 20% full and its owner adds 30 litres of petrol. If the tank is now 80% full what is the total capacity of the tank?

Answers pages 150–151

REVISION TEST 9 Level of difficulty—Average

1. There are 30 students in 7P. If 10% of the students are absent, how many are present?

2. Every payday three workmates contribute 1% of their pay to the cost of supporting a sponsor child. If they earn \$900, \$800 and \$1050 how much money is donated each payday?

3. Toni and Yvette are both paid \$720 each week. Toni uses one-third of her pay for her rent while Yvette uses 30%. Who pays more money in rent and by how much?

4. Tyler receives a 4% pay rise. If he was paid \$1200 per fortnight before the rise, what will be his new pay?

5. Meg spent \$120 on a pair of shoes and \$60 on a new wallet. If she had started with \$240 to spend, what percentage does she have remaining?

6. George bought his apartment four years ago for \$480 000. The value of the apartment has increased by 30%. What is its present value?

Answers page 151

REVISION TEST 10 Level of difficulty—Challenging

1. Martha and Mary are each given the same number of toffees to sell at the local fete. Martha sells 90% of her toffees while Mary only sells 70% of her toffees. Both girls sell their toffees for the same price. If they raise a total of $240 between them, how much more money has Martha made compared to Mary?

2. Barry played 80 games for his club. He scored at least one goal in 30% of the first 30 games he played. He scored in 40% of the next 40 games. In the remaining games he scored in 10% of the games. In how many games did Barry not score a goal?

3. A jewellery store increased the price of every item by the same percentage. The price of a watch increased from $400 to $480. What will be the new price of a bangle that was originally priced at $140?

4. A mixture of concentrate and water is used to fill a 12-litre container. It is known that 60% of the mixture is concentrate. Trent needs to add water to the mixture so that only 40% of the mixture is concentrate. What size container will he need for the new mixture?

5. There were 1060 students who attended a particular high school last year. Throughout the year, 60% of all boys and 40% of all girls represented the school in at least one sport. If 220 girls represented the school, how many boys represented the school?

6. A store sells leather jackets for $320. This price is 60% more than the store's cost price. At the end of winter, employees are able to buy any unsold stock at 20% under the store's cost price. What percentage discount off the store's original price do employees receive?

Answers pages 151–152

ALGEBRA

KEY SKILL

39 Using pronumerals

HINTS

- A pronumeral is a letter that is used to replace a number.
- Save time and space using notation: $1 \times a = a$, $b \times b = b^2$, $d \times c = c \times d = cd$, $e^1 = e$.
- When expressing a problem involving pronumerals, pretend the letters are numbers and see what happens,
  e.g. How many minutes in k hours?
  $60 \times k = 60k$ (If it was 3 hours, then it would be 60×3.)
  e.g. How much will p chocolates at q cents each cost?
  $p \times q = pq$ cents (If it was 3 chocolates at 10c each, it would be 3×10.)

Reminder!

- Read the question carefully to identify what needs to be found.
- Reread the question when you've finished your answer to make sure that the question has been answered and your solution makes sense.

Examples

At the local supermarket, boxes of biscuits are on special. Each box contains *b* biscuits. Quentin buys *p* boxes and Yvonne buys *q* boxes. How many biscuits are bought in total?

Solution

$$\text{Total biscuits} = p \times b + q \times b$$
$$= pb + qb$$

$\therefore$ a total of $(bp + bq)$ biscuits are bought.

FOCUS on...

1. The question: Asks you to find the number of biscuits.
2. The information: Gives the number of biscuits in each box and the number of boxes bought by two people.
3. Your working: When working out questions like these in terms of pronumerals, it might be easier to replace the letters with numbers to think through the solution. For example, if each box had 10 biscuits and you bought 3 boxes, then you would multiply to get a total of 30.
   In this question, if each box had b biscuits and Quentin bought p boxes then you just multiply: $p \times b$ gives pb, which can be written as bp (because we usually write the letters in alphabetical order). Also, $q \times b$ gives qb, or bq.
   To find the total, just add: $bp + bq$.
4. Your answer: Now write a final statement, making sure to write the units:
   The total is $(bp + bq)$ biscuits.

At the school fete, Carlos bought *k* cupcakes at *x* cents each and *p* cookies at *y* cents each. How much change did he receive from \$*a*?

Solution

$$\text{Change} = 100 \times a - (k \times x + p \times y)$$
$$= 100a - (kx + py)$$

$\therefore$ the change is $(100a - (kx + py))$ cents.

FOCUS on...

1. The question: Asks you to find the change.
2. The information: Gives the prices of cupcakes and cookies, how many of them were bought and the amount that was paid for the purchase.
3. Your working: When working out questions like these in terms of pronumerals, it might be easier to replace the letters with numbers to think through the solution. For example, if a cupcake cost 20 cents and you bought 3 of them, then you would multiply to get 60 cents.
   In this question, if a cupcake cost x cents and Carlos bought k of them then you would just multiply: $k \times x$ gives kx cents.
   Also, $p \times y$ gives py cents.
   Now to find the total, just add: $kx + py$ cents, which is written as $(kx + py)$ cents.
   Now, to change Carlos's \$$a$ to cents you have to multiply the number of dollars by 100: this means $100 \times a = 100a$, which is $100a$ cents.
   The change is found by subtracting:
   $100a - (kx + py)$.
4. Your answer: Now write a final statement, making sure to write the units:
   The change is $(100a - (kx + py))$ cents.

Now try these!

1. Leon buys p kilograms of king prawns. There were q prawns in every kilogram. Before he left the fish market he decided to go back and buy another r kilograms of the same prawns. How many prawns did he buy altogether?

2. Lena bought t folders at p cents each and w pens at b cents each. How much change will she receive from $\$q$?

3. A maths test has p questions. Each question is worth k marks and Glen gets q questions wrong. What was his total score?

4. Kevin is three times the age of his son Alvin. Lennie is Kevin's father and was d years old when Kevin was born. How old is Lennie if Alvin is b years old?

5. Robert has a mass of w kilograms. He is t kilograms heavier than Jack. Jack is y kilograms lighter than Max. What is Max's mass?

6. CHALLENGE A box contained x oranges. Ming found that one in three oranges in the box were rotten. After she threw the rotten oranges away she used y oranges to make some juice. How many oranges remained in the box?

Answers page 152

KEY SKILL

40 Equations A

HINTS

- An equation is an algebraic expression with an equals sign.
- When solving an equation, the aim is to finish with the pronumeral equal to a number.
- As you solve you must remember to add/subtract/multiply/divide both sides of the equation by the same amount at the same time.

Reminder!

- Read the question carefully to identify what needs to be found.
- Reread the question when you've finished your answer to make sure that the question has been answered and your solution makes sense.

Examples

Wendy is four years younger than her sister. The sum of their ages is 28. Solve an equation to find Wendy's age.

Solution

Let Wendy's age be x.

$\therefore$ Wendy sister's age is $x + 4$.

$$x + x + 4 = 28$$
$$2x + 4 = 28$$
$$2x + 4 - 4 = 28 - 4$$
$$\frac{2x}{2} = \frac{24}{2}$$
$$x = 12$$

$\therefore$ Wendy is 12 years old.

FOCUS on...

1. The question: Asks you to write and solve an equation to find Wendy's age.
2. The information: Gives the sum of the ages of two girls and how much younger Wendy is than her sister.
3. Your working: An equation has a pronumeral and so let x be Wendy's age. Her sister is 4 years older, so her age is x + 4.

   As the sum of their ages is 28 then write the equation:

   $x + x + 4 = 28$

   Simplify this line and write:

   $2x + 4 = 28$

   Subtract 4 from both sides of the equation:

   $2x + 4 - 4 = 28 - 4$

   $2x = 24$

   Divide both sides of the equation by 2:

   $\frac{2x}{2} = \frac{24}{2}$

   $x = 12$
4. Your answer: Now write a final statement, making sure to write the units:

   Wendy is 12 years old.

Three cans of soft drink and a bottle of water cost $10. The cost of a can of soft drink is 40 cents more than the cost of a bottle of water. Solve an equation to find the cost of a can of soft drink.

Solution

Let the cost of a can of soft drink be x.

$\therefore$ the cost of a bottle of water is $x - 40$.

$$3x + x - 40 = 1000$$
$$4x - 40 = 1000$$
$$4x - 40 + 40 = 1000 + 40$$
$$\frac{4x}{4} = \frac{1040}{4}$$
$$x = 260$$

$\therefore$ the cost of a can of soft drink is $2.60.

FOCUS on...

1. The question: Asks you to write and solve an equation to find the cost of a can of soft drink.
2. The information: Gives the total cost of three cans of soft drink and a bottle of water and how much more a soft drink costs compared to water.
3. Your working: Let x be the cost of a can of soft drink in cents. If this is 40 cents more than the water, then the water will cost (x – 40) cents.

   There are 3 cans which is 3x.

   The sum of 3 cans and a bottle is $10, or 1000 cents, so write the equation:

   $3x + x - 40 = 1000$

   This line can be rewritten as $4x - 40 = 1000$

   Add 40 to both sides of the equation:

   $4x - 40 + 40 = 1000 + 40$

   $4x = 1040$

   Now divide both sides of the equation by 4:

   $\frac{4x}{4} = \frac{1040}{4}$

   $x = 260$
4. Your answer: Now write a final statement, making sure to write the units:

   The can costs 260 cents, or $2.60.

Now try these!

1. Ken is five years older than his sister Meg. If the sum of their ages is 31 years, how old is Meg?

2. The cost of two apples and an orange is $1.70. If the cost of an orange is 20 cents more than an apple what is the cost of an apple?

3. Elton has two containers of rice. One container has 800 grams more rice than the other container. If the total mass of rice in both containers is 3.2 kg, how much rice is in the larger container?

4. Armir runs for twice as long as he swims each morning. If together he spends 75 minutes exercising, how long does he swim?

5. At a small high school there are two classes in Year Seven. One class has seven more students than the other class. Altogether there is a total of 49 Year-Seven students. How many students are in each class?

6. CHALLENGE The cost of a pen is 30 cents more than the cost of a pencil. Jackson spent $4.30 when he bought five pens and two pencils. What is the cost of each?

Answers pages 152–153

KEY SKILL

41 Equations B

HINTS

- Make sure you know the **Equations Hints** from Key Skill 40.
- Equations can be used in other areas of mathematics such as measurement, geometry, etc. You need to use rules relating to perimeter, area and volume formulae, angles, triangles, quadrilaterals, parallel lines, and so on.

Reminder!

- Read the question carefully to identify what needs to be found.
- Reread the question when you've finished your answer to make sure that the question has been answered and your solution makes sense.

Examples

The length of a rectangle is twice the width. The perimeter is 54 cm. Solve an equation to find the dimensions of the rectangle.

Solution

Let the width of the rectangle be x.

$\therefore$ the length of the rectangle is $2x$.

$$2 \times (x + 2x) = 54$$
$$2 \times 3x = 54$$
$$6x = 54$$
$$\frac{6x}{6} = \frac{54}{6}$$
$$x = 9$$

$\therefore$ the dimensions are 18 cm and 9 cm.

FOCUS on...

1. The question: Asks you to write and solve an equation to find the length and width.
2. The information: Gives the perimeter of a rectangle and the comparison of the length and the width.
3. Your working: Let x be the width of the rectangle. The length is 2x because it is twice the width.
   The perimeter of a rectangle is two times the sum of the length and the width:
   2 × (length + width) = perimeter
   Write what you know into the formula so that you can see what you have to find:
   $2 \times (x + 2x) = 54$
   You can add x and 2x to get 3x and then multiplying this by 2 gives 6x:
   $6x = 54$
   Now divide both sides of the equation by 6:
   $\frac{6x}{6} = \frac{54}{6}$
   $x = 9$
   The width was x, so the width is 9 and the length is 18.
4. Your answer: Now write a final statement, making sure to write the units: The dimensions are 18 cm and 9 cm.

The size of an angle in a triangle is 40° more than one angle and 10° less than the other angle. Solve an equation to find the size of each angle.

Solution

Let the middle-sized angle be $x°$.

$\therefore$ the other two angles are $(x - 40)°$ and $(x + 10)°$

$$\therefore x + x - 40 + x + 10 = 180$$
$$3x - 30 = 180$$
$$3x - 30 + 30 = 180 + 30$$
$$3x = 210$$
$$\frac{3x}{3} = \frac{210}{3}$$
$$x = 70$$

$\therefore$ the angles are 70°, 30°, 80°.

FOCUS on...

1. The question: Asks you to write and solve an equation to find the size of each of the angles of a triangle.
2. The information: Gives the comparison of the angles of a triangle and you know the sum is 180°.
3. Your working: Let x be the size of one of the angles. It is 40° more than another which you call (x − 40)°. It is also 10° less than the third angle which can be (x + 10)°.
   The angles in a triangle add to 180° so form an equation:
   $x + x - 40 + x + 10 = 180$
   This line is rewritten:
   $3x - 30 = 180$
   Now, add 30 to both sides of the equation:
   $3x - 30 + 30 = 180 + 30$
   $3x = 210$
   Divide both sides of the equation by 3:
   $\frac{3x}{3} = \frac{210}{3}$
   $x = 70$
   The angles are 70°, (70 − 40)° = 30° and (70 + 10)° = 80°.
4. Your answer: Now write a final statement, making sure to write the units: The angles are 70°, 30° and 80°.

Now try these!

1. The length of a rectangle is twice the breadth. If the perimeter of the rectangle is 72 cm, what are dimensions of the rectangle?

2. In a triangle the largest angle is 10° bigger than another angle and 50° bigger than the other angle. What is the size of each angle?

3. In a triangle, one side is three times as large as the smallest side and the third side is 10 cm more than the smallest side. The perimeter of the triangle is 45 cm. Find the dimensions of all three sides.

4. One angle of a triangle is twice the size of the smallest angle. The third angle is three times the size of the smallest angle. What is the size of each of the angles?

5. If one side of a square is increased by 4 cm and the other side decreased by 4 cm a rectangle is formed. If the rectangle has a perimeter of 24 cm, what were the dimensions of the original square?

6. CHALLENGE The length of a rectangle is 6 cm less than twice its width. If the perimeter of the rectangle is 36 cm, what is its area?

Answers pages 153–154

REVISION TEST 11 Level of difficulty—Average

1. Li shops at the fruit shop and buys p apples at t cents each and q oranges at v cents each. What is the total cost of the purchases?

2. Aretha is d years old. Her mother is 20 years older than her. When her mother was born Aretha's aunt was k years old. How old is Aretha's aunt?

3. Greg buys two hamburgers and a juice and pays $12. The price of a juice is $3 less than the price of a hamburger. Find the cost of a hamburger.

4. Three sisters have a total of $340. The oldest sister has $40 more than the middle sister and $10 more than the youngest sister. How much money has the youngest sister?

5. The largest angle in a triangle is 40° greater than the smallest angle. If the smallest angle is 20° less than the other angle, what are the sizes of all three angles?

6. Three friends had dinner at a restaurant and the total bill was $200. At the end of the night Keith put in $20 more than Rita but $10 less than Jane. How much did Rita pay?

Answers page 154

1. Simon's backyard pool is rectangular in shape with dimensions a metres by b metres. Surrounding the pool is a paved area which is d metres wide. What is the area of the paved section?

2. Melissa is twice as old as Harriet and Lionel is six years older than Harriet. Their ages total 42. How old will Lionel be when Melissa is 36?

3. There are 10 questions in a maths test. Each correct answer is given 5 marks but 2 marks are deducted for each incorrect answer. Edward answered every question and scored 72%. How many answers did he answer correctly?

4. For a school project Zelda has to test three brands of hand cream. Brand A costs twice as much as Brand B. Brand C costs $2.70 less than Brand A. Altogether Zelda spent $14.80 buying the hand creams. What is the cost of Brand C?

5. The length of a park's garden bed is 12 m longer than twice the width. If the perimeter of the garden is 144 m what is the cost of mulching the garden to a depth of 60 cm if mulch costs $0.20 per m^3?

6. Five years ago Garry was three times as old as Annie. In 12 years he will be two times as old as Annie. How old is Garry today?

Answers pages 154–155

KEY SKILL

42 Area A

HINTS

- The area is a measure of space inside a plane shape. The formulae are:

Rectangle	Square	Triangle	Parallelogram
length × width	side × side	$\frac{1}{2}$ × base × height	base × height

e.g. A rectangle with area 20 cm^2 is partly covered by a smaller rectangle with area 8 cm^2 which sits inside. What area of the larger rectangle is seen?

Area = 20 − 8
= 12 ∴ area seen is 12 cm^2.

Reminder!

- Read the question carefully to identify what needs to be found.
- Reread the question when you've finished your answer to make sure that the question has been answered and your solution makes sense.

Examples

Colin decided to plant a vegetable patch. He wants to make it 10 metres long and 6 metres wide. He builds a fence surrounding the garden. He decides to make the fence 12 metres long and 8 metres wide. What is the area between the vegetable patch and the fence?

Solution

Area = 12 × 8 − 10 × 6
= 96 − 60
= 36

∴ the area is 36 m^2.

FOCUS on ...

1. The question: Asks you to find the area between the two rectangles.
2. The information: Gives the length and width of two rectangles.
3. Your working: Find the area of the larger rectangle and subtract the area of the smaller rectangle:
   Area = 12 × 8 − 10 × 6
   = 96 − 60
   = 36

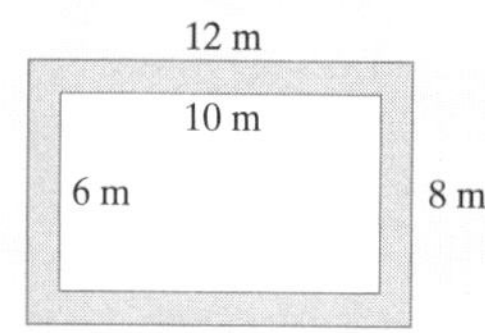

4. Your answer: Now write a final statement, making sure to write the units: The area is 36 m^2.

Fiona has sewn a rectangular patchwork quilt measuring 240 cm by 180 cm. Around the edge of the quilt she attached a satin border which was 20 cm wide. What is the area of the satin border?

Solution

Area = 280 × 220 − 240 × 180
= 61 600 − 43 200
= 18 400

∴ the area is 18 400 cm^2.

FOCUS on ...

1. The question: Asks you to find the area of the border.
2. The information: Gives the length and width of a rectangle and the size of the border around it.
3. Your working: The dimensions of the larger rectangle can be found by adding two lots of 20 cm to both the length and the width of the smaller rectangle.
   Area = 280 × 220 − 240 × 180
   = 61 600 − 43 200
   = 18 400

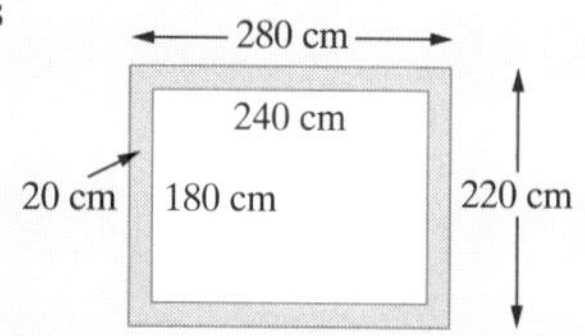

4. Your answer: Now write a final statement, making sure to write the units: The area is 18 400 cm^2.

Now try these!

1. Helen had a pool built in her backyard. The pool is rectangular with dimensions 9 metres by 4 metres. The fence surrounding the pool is 12 metres long and 6 metres wide. What is the area between the pool and the fence?

2. A photo is 32 cm by 20 cm. It is to be mounted on cardboard in a photo frame with a 4 cm border between the frame and the edge of the photo. What is the area of the cardboard that can be seen?

3. The top of a timber desk measures 180 cm by 120 cm. On the desk is a laptop which measures 50 cm by 34 cm. If there are no other items on the desk, what area of timber can be seen on the top of the desk?

4. Ray lays a path 1 metre wide around a rectangular garden. If the garden measures 8 metres by 6 metres, what is the area of the path?

5. Quentin has a corkboard in his bedroom. The corkboard is in the shape of a rectangle with a length of 120 cm and a width of 100 cm. He paints a border on the corkboard with an area of 660 cm^2. What is the size of the unpainted area?

6. CHALLENGE A backyard is in the shape of a parallelogram with a height of 8 m and a base length of 12 m. A square garden of side 6 metres has been built and the rest grassed. What is the area of the lawn?

Answers pages 155–156

KEY SKILL

43 Area B

HINTS

- Make sure you know the **Area Hints** from Key Skill 42.
- You can find the cost of fertilising, painting, covering, etc. by first finding an area and multiplying by a rate,
  e.g. If the area of a shape is 12 m^2, what is the cost of fertilising at \$2 per m^2?
  Cost = 12 × 2
  = 24 ∴ \$24

Reminder!

- Read the question carefully to identify what needs to be found.
- Reread the question when you've finished your answer to make sure that the question has been answered and your solution makes sense.

Examples

A wall measures 12 metres by 10 metres. Jack is to spraypaint the wall with paint that costs \$0.20 per square metre. What will be the cost of the paint?

Solution

Area = 12 × 10
= 120
∴ the area is 120 m^2.
Cost = 120 × 0.20
= 24
∴ the cost will be \$24.

FOCUS on …

1. The question: Asks you to find the cost of the paint used to paint a rectangular wall.
2. The information: Gives the length and width of the rectangle and the cost of the paint per m^2.
3. Your working: First, find the area of the field by multiplying the length by the width:
   12 × 10 = 120
   This means the area is 120 m^2.
   Then, multiply the number of square metres by the cost to paint each square metre: 120 × 0.20.
   In this question there are two digits that follow decimal places; this means the answer will have two digits that follow the decimal place.
   This means 120 × 0.**20** = 24.**00**
4. Your answer: Now write a final statement, making sure to write the units: The cost is \$24.

The council is returfing a field which measures 130 metres by 80 metres. If turf costs \$1.10 per square metre, what is the total cost of the returfing?

Solution

Area = 130 × 80
= 10 400
∴ the area is 10 400 m^2.
Cost = 10 400 × 1.10
= 11 440
∴ the cost is \$11 440.

FOCUS on …

1. The question: Asks you to find the cost of returfing.
2. The information: Gives the length and width of the rectangle and the cost of the turf per m^2.
3. Your working: First, find the area of the field by multiplying the length by the width:
   130 × 80 = 10 400
   This means the area is 10 400 m^2.
   Then multiply the number of square metres by the cost of each square metre: 10 400 × 1.10.
   In this question there are two digits that follow decimal places; this means the answer will have two digits that follow the decimal place.
   This means 10 400 × 1.**10** = 11 440.**00**
4. Your answer: Now write a final statement, making sure to write the units: The cost is \$11 440.

Now try these!

1. Lyle needs to buy a tarpaulin in the shape of a square with side length 9 metres. The cost of the material is $12 per square metre. What is the total cost?

2. A pair of stage curtains in a school hall is to be replaced. Each curtain measures 6 metres by 4 metres. The cost of the material is $40 per square metre. What is the total cost of the two curtains?

3. A rectangular paddock with dimensions 450 metres by 280 metres is to be fertilised. Find the total cost if the fertiliser costs 2 cents per square metre.

4. A yacht has a sail in the shape of a right-angled triangle. The base of the shape is 3.4 metres and the height is 5 metres. If the material in the sail costs $70 per square metre, what is the total cost of the material in the sail?

5. A farmer grows canola in a rectangular paddock with sides 120 metres by 210 metres. The canola is growing at an average of 78 plants per square metre. How many plants are in the paddock?

6. CHALLENGE A quilt is to be made by joining small squares with side length of 20 cm. A finished quilt measures 2 metres by 1.6 metres. Each small square is knitted from wool that costs $2.60. What is the cost of the wool in the completed quilt?

Answers pages 156–157

KEY SKILL

44 Area C

HINTS

- Make sure you know the **Area Hints** which appeared on previous pages.
- Sometimes the rate is expressed as a multiple unit instead of a single unit such as cost per 10 m^2, per 4 L, etc.,
  e.g. If the area of a shape is 50 m^2, what is the cost of covering it at \$3 per 10 m^2?
  Cost = 50 ÷ 10 × 3
  = 15 ∴ \$15

Reminder!

- Read the question carefully to identify what needs to be found.
- Reread the question when you've finished your answer to make sure that the question has been answered and your solution makes sense.

Examples

To maintain the temperature of his pool, Paul wants to use a pool cover measuring 5 metres by 12 metres. If the material used to manufacture the cover costs \$70 for every 4 square metres, what will be the cost of the material used in Paul's cover?

Solution

Area = 5 × 12
= 60
∴ the area of the pool cover is 60 m^2.
Cost = 60 ÷ 4 × 70
= 1050
∴ the cost will be \$1050.

FOCUS on...

1. The question: Asks you to find the cost of the pool cover.
2. The information: Gives the length and width of the rectangle and the cost of the material.
3. Your working: First, find the area of the cover by multiplying the length by the width:
   12 × 5 = 60
   This means the area is 60 m^2.
   Every 4 m^2 of cover costs \$70 so you need to find how many 4s are in 60. This is found by dividing:
   60 ÷ 4 = 30 ÷ 2 = 15
   The cover will cost 15 lots of \$70. Now, multiply 15 and 70:
   15 × 70 = 1050
4. Your answer: Now write a final statement, making sure to write the units: The cover will cost \$1050.

A large wall measuring 8 metres by 3 metres is to be painted. A litre of paint covers 4 m^2. If the paint is sold in 4-litre cans, how many cans are needed for two coats of paint?

Solution

Area = 8 × 3
= 24
Area of 2 coats = 24 × 2
= 48
∴ the area for 2 coats is 48 m^2.
Area for 4-L can = 4 × 4
= 16
∴ each can covers 16 m^2.
Number of cans = 48 ÷ 16
= 3
∴ 3 cans of paint are needed.

FOCUS on...

1. The question: Asks you to find the number of cans needed.
2. The information: Gives the length and width of the wall and the area covered by a litre of paint.
3. Your working: First, find the area of the wall by multiplying the length by the width: 8 × 3 = 24.
   This means the area is 24 m^2. As there are 2 coats, multiply the area by 2:
   24 × 2 = 48.
   An area of 48 m^2 will be painted.
   A litre of paint covers 4 m^2. Now, multiply to find the area covered by a 4-L can: 4 × 4 = 16.
   This means each tin covers 16 m^2.
   To find the number of cans of paint you need to divide the area of the wall by the area that can be covered by each can of paint: 48 ÷ 16 = 3.
4. Your answer: Now write a final statement, making sure to write the units: 3 cans of paint are needed.

Now try these!

1. Bill's Farm Fertiliser company charges $12 for every 1000 m^2 it fertilises. A paddock is 350 metres long and 200 metres wide. What will the company charge?

2. A deck measuring 12 metres by 8 metres is to be painted. If 0.5 litres of decking oil cover an area of 8 m^2, how many litres of the oil are required?

3. Mr Graham wants to cover a section of the school's gymnasium floor with mats. It is known that four mats cover a total area of 25 m^2. If the section of floor measures 30 m by 20 m, how many mats will he need?

4. Graeme has a triangular section of his property infested with a noxious weed. The section has a base of 40 metres and a height of 28 metres. If he controls the weed using a spray at the rate of 5 mL per 10 m^2, what quantity of the spray is needed?

5. Carpet tiles are square in shape and have a side length of 40 cm. The cost of each tile is $8. Eric uses the tiles to cover a rectangular floor measuring 3.6 metres by 4 metres. How much will it cost to cover the floor with tiles?

6. CHALLENGE 'OneCoat' ceiling paint is applied at the rate of 12 m^2 per litre. A 1-litre tin costs $21 and a 4-litre tin costs $56. Lee wants to paint a ceiling with dimensions 5.5 m by 5 m. What is the cheapest cost of the paint?

Answers page 157

KEY SKILL

45 Volume A

HINTS

- The volume is a measure of space inside a solid shape. The formulae are:

Rectangular prism	Cube
length × width × height	side × side × side

e.g. A rectangular prism has dimensions 5 cm by 4 cm by 3 cm. Find its volume.

Volume $= 5 \times 4 \times 3$

$= 60$ $\therefore$ volume is 60 cm^3.

Reminder!

- Read the question carefully to identify what needs to be found.
- Reread the question when you've finished your answer to make sure that the question has been answered and your solution makes sense.

Examples

Aaron digs a garden bed in his backyard. The dimensions of the bed are 4 metres by 3 metres by 80 centimetres. How much soil will he need to fill the garden bed?

Solution

Volume $= 4 \times 3 \times 0.8$

$= 9.6$

$\therefore$ the volume of the soil will be 9.6 m^3.

FOCUS on ...

1. The question: Asks you to find the volume of the garden bed.
2. The information: Gives the length, width and height of the rectangular prism. Dimensions are in metres and centimetres.
3. Your working: First, change all the dimensions to the same units: to keep the numbers small change them to metres. As there are 100 cm in a metre, then 80 cm is 0.8 m.
   Find the volume by multiplying the length by the width by the depth (height):
   $4 \times 3 \times 0.8 = 9.6$
   Remember, in the question there was one digit that followed a decimal point; this means the answer will have one digit following the decimal point.
4. Your answer: Now write a final statement, making sure to write the units: The volume is 9.6 m^3.

Alma's native garden has an area of 72 m^2. She wants to cover it with compost to a depth of 20 cm. What volume of compost will she need to order?

Solution

Volume $= 72 \times 0.2$

$= 14.4$

$\therefore$ the volume of the compost will be 14.4 m^3.

FOCUS on ...

1. The question: Asks you to find the volume of the compost.
2. The information: Gives the area and depth of the prism. Dimensions are in metres and centimetres.
3. Your working: First, change all the dimensions to the same units: to keep the numbers small change them to metres. As there are 100 cm in a metre, then 20 cm is 0.2 m.
   Find the volume by multiplying the area by the depth:
   $72 \times 0.2 = 14.4$
   Remember, in the question there was one digit that followed a decimal point; this means the answer will have one digit following the decimal point.
4. Your answer: Now write a final statement, making sure to write the units: The volume is 14.4 m^3.

Now try these!

1. Dylan needs to pour a concrete slab for a double garage in his backyard. The slab is 8 metres long, 6 metres wide and 0.6 metres deep. How much concrete is needed?

2. A chocolate bar is in the shape of a triangular prism. The area of its base is 6.2 cm^2. If the chocolate is 12 cm high what is its volume?

3. The dimensions of a brick paver are 22 cm long by 11 cm wide and 4 cm deep. What is the volume of the paver?

4. A length of metal bar has a cross-sectional area of 12 cm^2. If the length is 2 metres, what is the volume of metal in the bar, in cm^3?

5. A shipping container is in the shape of a square prism. The length of the container is 4 metres and its front is a square with side 2 metres. What is the volume of the container?

6. CHALLENGE A company buys lead ingots. The dimensions of each ingot are 600 mm by 100 mm by 100 mm. If the company melted an ingot to have a base of 200 mm by 200 mm, how high would be the ingot?

Answers page 157

KEY SKILL

46 Volume B

HINTS

- Make sure you know the **Volume Hints** from Key Skill 45.
- You can find the cost of filling, etc. by first finding the volume and multiplying by a rate,
  e.g. If the volume of a solid is 20 m^3, what is the cost of filling it at \$3 per m^3?
  Cost = 20 × 3
  = 60 ∴ \$60

Reminder!

- Read the question carefully to identify what needs to be found.
- Reread the question when you've finished your answer to make sure that the question has been answered and your solution makes sense.

Examples

A garden shop sells mushroom compost in boxes. Each box has a length of 40 cm, a width of 20 cm and a height of 10 cm. The mass of each cm^3 of compost is 0.4 grams. What is the mass of the compost in the box?

Solution

Volume = 40 × 20 × 10
= 8000
∴ the volume of the compost is 8000 cm^3.
Mass = 8000 × 0.4
= 3200
∴ the mass is 3200 grams, or 3.2 kg.

FOCUS on …

1. The question: Asks you to find the mass of compost.
2. The information: Gives the length, width and height of the rectangular prism and the mass of each cubic centimetre.
3. Your working: First, find the volume of the compost by multiplying the length by the width by the height:
   40 × 20 × 10 = 8000
   This means the volume is 8000 cm^3.
   To find the mass of the compost, multiply the total volume by the mass for each cm^3:
   8000 × 0.4 = 3200
4. Your answer: Now write a final statement, making sure to write the units: The mass is 3200 g or 3. 2 kg.

Mary has cooked a large lasagna for her dinner party. The lasagna measures 30 cm by 20 cm by 8 cm. If Mary divides it equally between 12 people, what volume does each person receive?

Solution

Volume = 30 × 20 × 8
= 4800
∴ the volume is 4800 cm^2.
Amount for each person = 4800 ÷ 12
= 400
∴ the volume for each person is 400 cm^3.

FOCUS on …

1. The question: Asks you to find the amount that each person receives.
2. The information: Gives the length, width and height of the rectangular prism and the number of people to share the lasagne.
3. Your working: First, find the volume of the lasagna by multiplying the length by the width by the height:
   30 × 20 × 8 = 4800.
   This means the volume is 4800 cm^3.
   To find the volume for each person, divide the total volume by the number of people:
   4800 ÷ 12 = 400
4. Your answer: Now write a final statement, making sure to write the units: The volume for each person is 400 cm^3.

Now try these!

1. A solid cube made of metal has a side length of 5 cm. Every cm^3 of the metal has a mass of 60 grams. What is the total mass of the cube in kg?

2. Rashid is baking a chocolate cake which is 30 cm wide and 20 cm long. The height of the cake is 10 cm. If the cake is to be shared equally between 15 people, what volume of cake will each person receive?

3. Mrs Carter divides a rectangular prism of clay equally between the 20 students in her ceramics class. The dimensions of the clay are 16 cm by 10 cm by 8 cm. What volume does each student receive?

4. A truck dumps a load of garden soil in Goran's front yard. The dimensions of the truckload are 4 metres long by 2 metres wide by 1.6 metres high. If the load is to be equally shared by Goran and three of his neighbours, what volume will each receive?

5. James uses a metal box to store seeds. The dimensions of the box are 20 cm by 10 cm by 8 cm. The mass of seeds is 2 g per cm^3. What is the mass of seeds in the box if the box is completely full of seeds?

6. CHALLENGE A box of cereal is full. It is 25 cm high by 20 cm wide by 6 cm deep. The total mass of the cereal in the box is 750 grams. What would be the mass of the cereal in a box with dimensions 20 cm by 15 cm by 8 cm?

Answers page 158

REVISION TEST 13 Level of difficulty—Average

1. A garden bed measures 12 metres by 10 metres. The garden has a 1-metre wide lawn border surrounding it. What is the area of the grass border?

2. Owen purchases an area of canvas which is shaped as a right-angled triangle. The base of the triangle is 4.2 metres and its height is 3 metres. If the cost of the canvas is $8 per square metre, what is the cost of Owen's canvas?

3. Willie wants to make a feature of one of his bedroom walls. The wall measures 3 metres by 2.8 metres. He uses gold paint to paint a parallelogram with length 2 metres and height 1.5 metres. The section outside the parallelogram is painted light blue. Find the area which is not painted gold.

4. To prepare for Canberra's Floriade, workers build a raised garden bed by placing soil to a depth of 0.3 metres over a rectangular area measuring 30 metres by 10 metres. How much soil is used to make the bed?

5. Arissa uses paint that covers at the rate of 4 m^2 per litre. How many litres are required to paint a square wall with two coats if the side length is 3 metres?

6. One of the faces of a cube has an area of 9 cm^2. What is the volume of the cube?

Answers pages 158–159

REVISION TEST 14 Level of difficulty—Challenging

1. The area of a parallelogram is 32 cm^2 and the length of its base is 6.4 cm. What is the length of its perpendicular height?

2. Harry used a 60-cm length of wire to make the frame of a cube. The cube was then enclosed by plastic. What is the volume of the cube?

3. The length of a rectangular prism is three times its width and the height is twice the width. If the volume of the shape is 48 cm^3 what is the height of the prism?

4. The cost of fertilising a rectangular-shaped paddock is $1440. The width of the paddock is 120 metres. If the cost of the fertiliser is $4 per 100 m^2, what is the length of the paddock?

5. Penny has boxes measuring 50 cm by 25 cm by 20 cm. How many boxes can she fit in a carton measuring 1 m by 1 m by 1 m?

6. Glen's trailer measures 1.6 m by 2 metres. He filled it with soil to a depth of 0.6 metres. He used some of the soil to fill a square garden bed with side 2 metres to a depth of 30 cm. What percentage of his trailer load remains?

Answers page 159

KEY SKILL

47 Averages A

HINTS

- The mean (average) is found by totalling the scores and dividing by the number of scores,
  e.g. What is the average of 3, 6, 8, 3?
  Average $= (3 + 6 + 8 + 3) \div 4$
  $= 20 \div 4$
  $= 5$
  $\therefore$ the average of the four scores is 5.
- If you know the average of a set of scores then you can find the total of the scores,
  e.g. If the average of six scores is 4, what is the total of the scores?
  Sum $= 4 \times 6$
  $= 24$
  $\therefore$ the sum of the six scores is 24.

Reminder!

- Read the question carefully to identify what needs to be found.
- Reread the question when you've finished your answer to make sure that the question has been answered and your solution makes sense.

Examples

The minimum temperature in Penrith on four consecutive days was 5°, 7°, 8° and 4°. What was the average minimum temperature across the four days?

Solution

Average $= (5 + 7 + 8 + 4) \div 4$
$= 24 \div 4$
$= 6$
$\therefore$ the average minimum temperature was 6°.

FOCUS on...

1. The question: Asks you to find the average minimum temperature across the four days.
2. The information: Gives the temperature on four days.
3. Your working: To find the average add the numbers then divide by how many numbers there are:
   $\frac{5 + 7 + 8 + 4}{4} = \frac{24}{4}$
   $= 6$
4. Your answer: Now write a final statement, making sure to write the units: The average was 6°.

Leith's average score in three maths tests was 16. His average score in two science tests was 11. What was the average across all five tests?

Solution

Average of 3 maths tests $= 16$
Total of 3 maths tests $= 16 \times 3$
$= 48$
Average of 2 science tests $= 11$
Total of 2 science tests $= 11 \times 2$
$= 22$
Total of the 5 tests $= 48 + 22$
$= 70$
Average of the 5 tests $= 70 \div 5$
$= 14$
$\therefore$ Leith's average mark was 14.

FOCUS on...

1. The question: Asks you to find Leith's average mark in his five tests.
2. The information: Gives the average of three tests and the average of another two tests.
3. Your working: If the average of his three maths tests was 16 then find the total of these marks by multiplying: $16 \times 3 = 48$.
   If the average of his two science tests was 11 then find the total of these marks by multiplying: $11 \times 2 = 22$
   Leith did three maths and two science tests. This means he has five test results.
   To find the average you need to add the totals and then divide by 5:
   $\frac{48 + 22}{5} = \frac{70}{5}$
   $= 14$
4. Your answer: Now write a final statement, making sure to write the units:
   The average mark for all tests was 14.

Now try these!

1. John cycled on five mornings. His distances were 5 km, 4.2 km, 3.7 km, 4.6 km and 3 km. What was the average distance he cycled across the days?

2. In the first four weeks Miriam worked she was paid an average of $60. In the next two weeks she was paid an average $90. What was the average amount she was paid for the six weeks' work?

3. It rained for the first five days of Brian's school holidays. He recorded the amounts as 16 mm, 8 mm, 10 mm, 5 mm and 16 mm. What was the average daily measurement?

4. In the first three months of the year Leo read an average of two books each month. In the following three months he read an average of four books each month. What was the average number of books read over the six months?

5. Kevin scored the following points for his basketball team: 18, 12, 15, 0, 16 and 23. What was the average number of points scored per game?

6. CHALLENGE At a cattle sale a farmer sold four steers at an average price of $650 each and six heifers at an average price of $620 each. What was the average price for all the cattle sold?

Answers pages 159–160

KEY SKILL

48 Averages B

HINTS

- Make sure you know the **Statistics Hints** from Key Skill 47.
- To find how the average changes with the inclusion or removal of a score you first need to find the way the total has changed.

Reminder!

- Read the question carefully to identify what needs to be found.
- Reread the question when you've finished your answer to make sure that the question has been answered and your solution makes sense.

Examples

Jason scored an average of 92 on three maths tests. What mark does he need on his next test to bring his average up to 94?

Solution

Average of 3 tests = 92
Sum of 3 tests = 92 × 3
= 276
Average of 4 tests = 94
Sum of 4 tests = 94 × 4
= 376
4th test result = 376 − 276
= 100
∴ Jason needs 100 in his next test.

FOCUS on ...

1. The question: Asks you to find Jason's result in his next test.
2. The information: Gives the average of three tests and then the average of four tests.
3. Your working: If the average of three tests is 92 then multiply to get the total of the three tests:
   Total of 3 tests = 92 × 3 = 276
   If the average of 4 tests is 94 then multiply to get the total of the 4 tests:
   Total of 4 tests = 94 × 4 = 376
   To find Jason's fourth test result, subtract the two totals:
   Fourth test = 376 − 276 = 100
4. Your answer: Now write a final statement:
   Jason needs 100 in his next test.

The average age of 10 people at a party is 21. When Peter leaves the party the average drops by 2. How old is Peter?

Solution

Average age of 10 people = 21
Total age of 10 people = 21 × 10
= 210
Total age of 9 people = 19 × 9
= 171
Peter's age = 210 − 171
= 39
∴ Peter is 39 years old.

FOCUS on ...

1. The question: Asks you to find Peter's age.
2. The information: Gives the average age of 10 people and then the average age of nine people.
3. Your working: If the average age of 10 people is 21 then multiply to get the total of the 10 ages:
   Total of 10 ages = 21 × 10 = 210
   If the average drops by 2, then subtract:
   New average = 21 − 2 = 19
   If the average age of nine people is 19, then multiply to get the total of the nine ages:
   Total of 9 ages = 19 × 9 = 171
   To find Peter's age, subtract the two totals:
   Peter's age = 210 − 171 = 39
4. Your answer: Now write a final statement:
   Peter is 39.

Now try these!

1. Dave's average on four science tests is 16 out of 25. What must he score on his next science test to increase the average to 17?

2. The average mass of six people in a group is 60 kg. If Karen leaves the group the average drops to 58 kg. What is her mass?

3. A cricketer has an average score of 24 in five innings. After his next innings his average has increased by four. What did he score in his sixth innings?

4. The average length of three straws is 8 cm. If the average of two of the straws is 10 cm, how long is the third straw?

5. The average height of the 'starting five' in a school's basketball team is 182 cm. If the height of one of the players is 186 cm, what is the average height of the other four players?

6. CHALLENGE There are six children in the Clarke family. The average age of the four boys is 12. One of the girls is 5. If the average age of the six children is 10, how old is the other girl?

Answers page 160

KEY SKILL

49 Diagrams, tables and guess-and-check A

HINTS

- Drawing a diagram can be helpful in summarising a problem and finding the solution.
- A table can be used to keep track of different possible solutions.
- Some problems are so complex that the student makes a guess at the answer and then checks its accuracy. A second, more educated, guess can then be made. These guesses should be recorded as a list or in a table as it is important to record your working.

Reminder!

- Read the question carefully to identify what needs to be found.
- Reread the question when you've finished your answer to make sure that the question has been answered and your solution makes sense.

Examples

Jack must transport a chicken, a fox and a bag of wheat across a river by boat. The boat can only hold the farmer and one of the objects. Unfortunately, if left alone, the fox will eat the chicken or the chicken will eat the wheat. How many times does the boat cross the river to manage it?

Solution

Jack takes the chicken across, then comes back in the boat by himself. He then takes the wheat across and then picks up the chicken and takes it back. Now Jack puts the fox in the boat and takes it across so that the fox is with the wheat. Jack finally goes back by himself and picks up the chicken to bring it across the river.

∴ Jack crosses the river seven times.

FOCUS on ...

1. The question: Asks you to find the number of river crossings.
2. The information: Gives the problem of three items to be taken across a river.
3. Your working: Use a diagram and letters to help explain the solution (J: Jack, c: chicken, w: wheat and f: fox).
   One side ... river ... other side
4. Your answer: Now write a final statement:
   The boat crosses the river seven times.

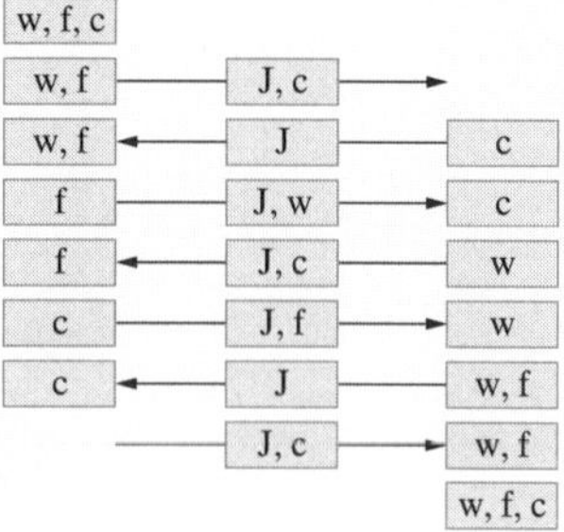

McDonald has cows and ducks on his farm. His animals have a total of 24 heads and 60 feet. How many cows and ducks live on McDonald's farm?

Solution

The number of cows and ducks adds to 24. If he has 6 cows (24 legs) and 18 ducks (36 legs) then he has a total of 24 heads and 60 legs.

FOCUS on ...

1. The question: Asks you to find the number of cows and ducks on the farm.
2. The information: Gives the number of heads and feet on the animals. The animals have one head, but cows are four-legged and ducks are two-legged.
3. Your working: A guess of 12 cows and 12 ducks has a total of 24 heads but 72 feet.
   Keep the number of heads by dropping the number of cows and increasing the number of ducks:

Guesses	Cows (4)	Ducks (2)	Heads	Feet
1	12	12	24	48 + 24 = 72
2	10	14	24	40 + 28 = 68
3	8	16	24	32 + 32 = 64
4	6	18	24	24 + 36 = 60

4. Your answer: Now write a final statement:
   On the farm there are six cows and 18 ducks.

Now try these!

1. A father and two sons need to cross a river in a canoe. The man has a mass of 80 kg and the boys have a mass of 40 kg each. The canoe will only hold a mass of 100 kg. What is the smallest number of trips across the river?

2. Darryl makes three-legged stools and four-legged stools. In one week he used 94 legs to make 26 stools. How many four-legged stools did he make?

3. Three cannibals and three missionaries must cross a river. The boat can only hold two people. The cannibals can never outnumber the missionaries on either side of the river. What is the smallest number of trips across the river?

4. Miriam is collecting spiders and beetles. Each spider has eight legs and each beetle has six legs. In her collection she has 10 creatures and a total of 64 legs. How many of each type has she collected?

5. Jackson has a sum of money made up of five-cent coins and ten-cent coins. There are 40 coins altogether and the money has a total value of $3.20. How many ten-cent coins has Jackson?

6. CHALLENGE Farm animals are available for sale in groups of five. Dave bought 100 animals and spent exactly $1000. He bought pigs, chickens and sheep. Each pig cost $20, each chicken $5 and each sheep $30. How many sheep did Dave buy?

Answers pages 160–161

KEY SKILL

50 Diagrams, tables and guess-and-check B

HINTS

- Read the question carefully to discover what needs to be found.
- Make sure you know the **Diagrams, tables and guess-and-check Hints** from Key Skill 49.

Reminder!

- Read the question carefully to identify what needs to be found.
- Reread the question when you've finished your answer to make sure that the question has been answered and your solution makes sense.

Examples

If the sum of four times a number and the number divided by 4 is decreased by 15, the result is 36. What is the number?

Solution

Try 12: $(4 \times 12 + 12 \div 4) - 15$
$= 36$

$\therefore$ the number is 12.

FOCUS on ...

1. The question: Asks you to find the number.
2. The information: Gives the result of a number sentence involving the number and various operations.
3. Your working: Use a guess-and-check method to find the number. First guess is 4: multiplying by 4 gives 16 and then dividing by 4 gives 1. Adding 16 and 1 gives 17. When you subtract 15 you have 2. This is not 36 so make further guesses.

Guessed	4 ×	÷ 4	Sum	Sum − 15
4	16	1	17	2
8	32	2	34	19
12	48	3	51	36

4. Your answer: Now write a final statement:
The number is 12.

Al's bakery makes chocolate, banana, blueberry and sultana scones. Three-eighths of the scones baked were chocolate. The number of banana scones was two-thirds the number of sultana scones. The number of blueberry scones was three-quarters the number of sultana scones. How many chocolate scones were baked if Al baked 45 blueberry scones?

Solution

Blueberry scones = 45
Sultana scones = $45 \div 3 \times 4 = 60$
Banana scones = $60 \div 3 \times 2 = 40$
Total of blueberry, sultana, banana
$= 45 + 60 + 40$
$= 145$

$\frac{5}{8}$ of total scones $= 145$

$\frac{3}{8}$ of total scones $= 145 \div 5 \times 3$
$= 87$

$\therefore$ there were 87 chocolate scones baked.

FOCUS on ...

1. The question: Asks you to find the number of chocolate scones baked.
2. The information: Gives the list of different scones and the fractions of each type.
3. Your working: A table is used to summarise the information in the question to make it easier to work through: There are 45 blueberry scones:

chocolate	
banana	
blueberry	45
sultana	

Number of sultana scones
$= 45 \div 3 \times 4$
$= 60$

chocolate	
banana	
blueberry	45
sultana	60

Number of banana scones
$= 60 \div 3 \times 2$
$= 40$

chocolate	
banana	40
blueberry	45
sultana	60

Total so far $= 40 + 45 + 60$
$= 145$

If three-eighths are chocolate, then five-eighths must be the rest.

Five-eighths of total= 145
One-eighth of total $= 145 \div 5$
$= 29$
Three-eighths of total $= 29 \times 3$
$= 87$

chocolate	87
banana	30
blueberry	60
sultana	45

4. Your answer: Now write a final statement:
Al baked 87 chocolate scones.

Now try these!

1. If the sum of three times a number and the number divided by 3 is increased by 9, the result is 59. What is the number?

2. In a lolly shop there are 60 more red frogs than green frogs. The number of green frogs is twice as many as the chocolate frogs and four times as many as the liquorice frogs. If there are 12 liquorice frogs how many red frogs are in the shop?

3. Janeel is twice as old as Kai and four times older than Penny. Owen is two years younger than Penny. How old is Janeel if the total of their ages is 70?

4. Divide $220 between Anne, Bennie and Sam so that Anne has $20 more than Bennie but $30 less than Sam. What is Sam's share?

5. Barry has three bills to pay. The first bill is three times the amount of the second one. The third bill is ten times the amount of the second bill. The difference between the first bill and the third bill is $1050. What is the total of the three bills?

6. CHALLENGE Jamie was given some money for her birthday. She spent $5 less than one-third of her money on a top and $15 more than half of the original amount to buy a skirt. If she had $4 left, how much money was she given as a present?

Answers pages 161–162

REVISION TEST 15 Level of difficulty—Average

1. Adele runs the following distances on five mornings: 3 km, 2.5 km, 1.8 km, 2.1 km and 0.6 km. What was the average (mean) distance she ran?

2. John scored an average of 84% on four of his tests. He wants to increase his average to 87%. What must he score on the next test to reach his goal?

3. A group of eight students measured their heights. The average of the three smaller students was 152 cm and the average of the remaining students was 164 cm. What was the average height of all the students?

4. If the sum of three times a number and 18 is decreased by 12, the result is 30. What is the number?

5. A farmer asks his daughter to go into the barn and count the number of chickens and pigs. She comes back and says she counted 56 legs. The farmer then sent his son into the barn who came back to say he had counted 20 heads. How many chickens were in the barn?

6. Henry, Harold and Horatio worked on a project and recorded the number of hours worked. Harold worked twice as long as Horatio and six hours longer than Henry. If the three men worked a total of 54 hours, how many hours were worked by Horatio?

Answers page 162

REVISION TEST 16 Level of difficulty—Challenging

1. The average of eight scores is 12. If two more scores are included the mean drops by 2. If one of these new scores is 3, what is the other new score?

2. Penny and Murray have a race to the top of a set of 40 stairs. Murray gives Penny a 17-step start. Penny is able to take two steps at a time while Murray takes three steps at a time. If they step at the same instant who will reach the top first?

3. Ken, Jen and their dog Ben are weighed in pairs. Ken and Jen have a total mass of 90 kg, Ken and Ben's mass is 60 kg and Jen and Ben's total mass is 50 kg. What is Ken's mass?

4. A frog is at the bottom of a 9-metre deep well. Each day it crawls up 4 metres and at night slips back 2 metres. How many days will it take for the frog to get out of the well?

5. Four people come to a river at night. There is a narrow bridge but it can only hold two people at a time. Because it is night, the torch has to be used when crossing the bridge. Angela can cross the bridge in 1 minute, her brother Brock in 2 minutes, her father Colin in 5 minutes and her grandmother Dorrie in 8 minutes. When two people cross the bridge they must move at the slower person's pace. What is the fastest time in which the four people can all cross the bridge?

6. Every hour a train leaves Sydney and travels to Newcastle. At the same time a train leaves Newcastle and travels to Sydney. The trains all travel at the same speed and always take 2 hours to complete the journey. If James catches the 9 am train from Newcastle, how many trains will he meet on the journey?

Answers pages 162–163

Worked solutions

NUMBER

Key Skill  Addition (pages 10–11)

1. **42 minutes**
   Time = 14 + 6 + 22
   = 42
   ∴ Connor takes 42 minutes.

2. **54 994**
   $$\begin{array}{r} 21\,122 \\ 10\,380 \\ +\ 23\,492 \\ \hline 54\,994 \end{array}$$
   ∴ the total population is 54 994.

3. **63**
   Sam's age = 2
   Sam's father's age = 2 + 28
   = 30
   Sam's grandfather's age = 30 + 33
   = 63
   ∴ Sam's grandfather is 63 years old.

4. **323 m**
   Height of Eureka Tower = 297
   Height of Sydney Tower = 297 + 12
   = 297 + 3 + 9
   = 309
   Height of Q1 = 309 + 14
   = 323
   ∴ Q1 is 323 metres high.

5. **274**
   $$\begin{array}{r} 69 \\ 70 \\ 68 \\ +\ 67 \\ \hline 274 \end{array}$$
   ∴ Ryan's total was 274.

6. **84**
   6 + 8 + 10 + 12 + 14 + 16 + 18 = 84
   ∴ Zac has completed 84 push-ups.

Key Skill  Subtraction (pages 12–13)

1. **368 000**
   Consider the number of thousands:
   $$\begin{array}{r} 547 \\ -\ 179 \\ \hline 368 \end{array}$$
   ∴ the difference is 368 000.

2. **1890 m**
   9670 − 3600 − 4180
   $$\begin{array}{r} 9670 \\ -\ 3600 \\ \hline 6070 \end{array} \qquad \begin{array}{r} 6070 \\ -\ 4180 \\ \hline 1890 \end{array}$$
   ∴ the plane is at 1890 m above sea level.

3. **600**
   $$\begin{array}{r} 9000 \\ -\ 3800 \\ \hline 5200 \end{array} \qquad \begin{array}{r} 5200 \\ -\ 4600 \\ \hline 600 \end{array}$$
   ∴ 600 books are still to be delivered.

4. **238 000 000**
   $$\begin{array}{r} 312 \\ -\ 74 \\ \hline 238 \end{array}$$
   ∴ Indonesia's population was 238 000 000.

5. **1983 km**
   $$\begin{array}{r} 2793 \\ -\ 435 \\ \hline 2358 \end{array} \qquad \begin{array}{r} 2358 \\ -\ 375 \\ \hline 1983 \end{array}$$
   ∴ Hannah still has to travel 1983 km.

6. **$998 999 000**
   $$\begin{array}{r} 1\,000\,000\,000 \\ -\ 1\,000\,000 \\ \hline 999\,000\,000 \end{array} \qquad \begin{array}{r} 999\,000\,000 \\ -\ 1\,000 \\ \hline 998\,999\,000 \end{array}$$
   ∴ $998 999 000 remains.

Key Skill  Multiplication (pages 14–15)

1. **$552**
   Total amount = 23 × 4 × 6
   = 92 × 6
   = 552
   ∴ the total paid is $552.

2. **$156**
   Total cost = 52 × 3
   = 156
   ∴ the total cost is $156.

3. **960**
   Total players = 16 × 10 × 6
   = 160 × 6
   = 960
   ∴ there are 960 players.

4. **$159**
   Cost = $5.30 × 5 × 6
   = $26.50 × 6
   = $159
   ∴ the cost is $159.

5. **3840**

Total seats $= 24 \times 20 \times 8$
$= 480 \times 8$
$= 3840$

$\therefore$ there are 3840 seats.

6. **201 600**

Total bottles $= 24 \times 20 \times 12 \times 7 \times 5$
$= 480 \times 84 \times 5$
$= 40\,320 \times 5$
$= 201\,600$

$\therefore$ there are 201 600 bottles delivered.

Key Skill  Division with remainders (pages 16–17)

1. **4**

Number of bags $= 29 \div 8$
$= 3\frac{5}{8}$

$\therefore$ Lauren needs to buy 4 bags.

2. **8**

Number of chairlifts $= 43 \div 6$
$= 7\frac{1}{6}$

$\therefore$ 8 chairlifts are needed.

3. **4**

Number of buses $= 180 \div 50$
$= 18 \div 5$
$= 3\frac{3}{5}$

$\therefore$ 4 buses need to be hired.

4. **14**

Number of buggies $= 53 \div 4$
$= 13\frac{1}{4}$

$\therefore$ 14 golf buggies were needed.

5. **542**

$$12\overline{)6500}\quad = 541\tfrac{8}{12}$$

$\therefore$ 542 cartons of eggs need to be purchased.

6. **$290**

As $40 \div 30 = 1\frac{1}{3}$, then Steele needs two 30-metre lengths, plus five 100-metre lengths.

Cost $= 2 \times \$20 + 5 \times \$50$
$= \$40 + \$250$
$= \$290$

$\therefore$ the cable will cost $290.

Key Skill  Addition and subtraction (pages 18–19)

1. **10 L**

Amount remaining $= 20 - (3 + 5 + 2)$
$= 20 - 10$
$= 10$

$\therefore$ there is 10 L remaining.

2. **220 km**

Distance $= 1300 - (380 + 410 + 290)$
$= 1300 - 1080$
$= 220$

$\therefore$ Andrew travelled 220 km on Monday.

3. **21**

Balls remaining $= 40 - (6 + 5 + 8)$
$= 40 - 19$
$= 21$

$\therefore$ Chris has 21 balls left.

4. **59**

Number of pedestrians
$= 453 - (210 + 102 + 69 + 13)$
$= 453 - 394$
$= 59$

$\therefore$ 59 pedestrians died.

5. **$75**

Remaining amount $= 250 - (40 + 65 + 70)$
$= 250 - 175$
$= 75$

$\therefore$ Taylah has $75 remaining.

6. **2 km**

Distance short $= 30 - (1 + 2 + 3 + 4 + 5 + 6 + 7)$
$= 30 - 28$
$= 2$

$\therefore$ Jason will be short by 2 km.

Key Skill  Addition and multiplication (pages 20–21)

1. **$13**

Total cost $= \$7 + 3 \times \$2$
$= \$7 + \$6$
$= \$13$

$\therefore$ the cost was $13.

2. **$186**

Total cost $= 8 \times \$12 + 5 \times \$18$
$= \$96 + \$90$
$= \$186$

$\therefore$ the total cost was $186.

3. **$185**
Total wage = $12 × 9 + $11 × 7
= $108 + $77
= $185
∴ their total wage is $185.

4. **136 years**
Total age = 12 × 7 + 13 × 4
= 84 + 52
= 136
∴ the total of their ages is 136 years.

5. **2400 m**
Total distance = 50 × (8 × 3 + 12 × 2)
= 50 × (24 + 24)
= 50 × 48
= 100 × 24
= 2400
∴ Allan swims 2400 m (or 2.4 km).

6. **$444 000**
Total = 120 000 + 72 000 × 2 + 28 000 × 6 + 12 000
= 120 000 + 144 000 + 168 000 + 12 000
= 444 000
∴ the total prizemoney was $444 000.

Key Skill  7 Addition and division (pages 22–23)

1. **19 kg**
Average = (25 + 14 + 18) ÷ 3
= 57 ÷ 3
= 19
∴ the new mass was 19 kg.

2. **28**
Average = $\frac{30 + 28 + 25 + 30 + 27 + 28}{6}$
= $\frac{168}{6}$
= 28
∴ the average class size is 28.

3. **$84**
Amount = (76 + 92) ÷ 2
= 168 ÷ 2
= 84
∴ they both end up with $84.

4. **$89**
Amount = (86 + 47 + 39 + 42 + 53) ÷ 3
= 267 ÷ 3
= 89
∴ each village will be given $89.

5. **22**
Average = $\frac{23 + 16 + 31 + 18}{4}$
= $\frac{88}{4}$
= 22
∴ the average was 22 goals.

6. **6**
Number = (13 + 16 + 10 + 18) ÷ 3
= 57 ÷ 3
= 19
Each player will have 19 balls.
∴ Janine receives 6 balls.

Key Skill  8 Multiplication and division (pages 24–25)

1. **20 days**
Number of days = 120 ÷ (3 × 2)
= 120 ÷ 6
= 20
∴ the hay lasts for 20 days.

2. **5 weeks**
Number of weeks = 60 ÷ (4 × 3)
= 60 ÷ 12
= 5
∴ it will take 5 weeks.

3. **15 days**
Number of days = 300 ÷ (10 × 2)
= 300 ÷ 20
= 30 ÷ 2
= 15
∴ the bottle will last 15 days.

4. **10 weeks**
Number of weeks = 840 ÷ (14 × 6)
= 840 ÷ 84
= 10
∴ it will take 10 weeks.

5. **6 days**
Number of days = 1440 ÷ (30 × 8)
= 1440 ÷ 240
= 144 ÷ 24
= 72 ÷ 12
= 6
∴ it will take 6 days.

6. **12**
Number = 28 800 ÷ (20 × 12 × 10)
= 28 800 ÷ 2400
= 288 ÷ 24
= 144 ÷ 12
= 12
∴ 12 trucks will be loaded.

Key Skill  Various operations

(pages 26–27)

1. **\$26**

$\text{Total cost} = \$3 \times 4 + \$6 \times 2$
$= \$12 + \$12$
$= \$24$
$\text{Change} = \$50 - \$24$
$= \$26$
$\therefore$ the change is \$26.

2. **\$1100**

$$\begin{array}{r} 420 \\ 380 \\ 375 \\ 410 \\ +\ 415 \\ \hline 2000 \end{array}$$

$\text{Number of bricks} = 2000$
$\text{Payment} = 2000 \div 1000 \times \$550$
$= 2 \times \$550$
$= \$1100$
$\therefore$ Ahmed is paid \$1100.

3. **\$105**

$\text{Cost of watch} = 450 - (85 \times 3 + 45 \times 2)$
$= 450 - (255 + 90)$
$= 450 - 345$
$= 105$
$\therefore$ the watch cost \$105.

4. **\$1260**

$$\begin{array}{r} 16 \\ 32 \\ 28 \\ 19 \\ 18 \\ +\ 27 \\ \hline 140 \end{array}$$

$\text{Payment} = 140 \div 10 \times 90$
$= 14 \times 90$
$= 1260$
$\therefore$ Bill is paid \$1260.

5. **1150 mL**

$2 \text{ litres} = 2000 \text{ mL}$
$\text{Remains} = 2000 - (300 \times 2 + 250)$
$= 2000 - (600 + 250)$
$= 2000 - 850$
$= 1150$
$\therefore$ 1150 mL remains in the carton.

6. **\$108**

$\text{Total distance} = 440 + 360$
$= 800$
$\text{Amount of petrol} = (800 \div 100) \times 9$
$= 8 \times 9$
$= 72$
$\text{Cost} = 72 \times \$1.50$
$= 36 \times \$3$
$= \$108$
$\therefore$ the cost of petrol for the weekend was \$108.

Revision Test 1
Level of difficulty—Average

(page 28)

1. **475**

$\text{Total} = 78 + 111 + 93 + 87 + 106$
$= 475$
$\therefore$ the total scored was 475 points.

2. **11**

$\text{Buses} = 465 \div 45$
$= 10$ and remainder 15
(as $45 \times 10 = 450$, and remainder 15)
$\therefore$ the school needs 11 buses.

3. **4915 L**

$\text{Used} = 80 + 120 + 175 + 105 + 205$
$= 685$
$\text{Remaining} = 5600 - 685$
$= 4915$
$\therefore$ 4915 L remains in the tank.

4. **\$22**

$\text{Change} = \$150 - (\$14 \times 4 + \$12 \times 6)$
$= \$150 - (\$56 + \$72)$
$= \$150 - \$128$
$= \$22$
$\therefore$ the change is \$22.

5. **\$120**

$\text{Average} = \dfrac{120 + 145 + 105 + 110}{4}$
$= \dfrac{480}{4}$
$= 120$
$\therefore$ Jana's average income is \$120.

6. **15 weeks**

$\text{Distance each week} = 15 \times 6$
$= 90$
$\text{No. of weeks} = 1350 \div 90$
$= 135 \div 9$
$= 15$
$\therefore$ it will take Tarek 15 weeks.

Revision Test 2
Level of difficulty—Challenging

(page 29)

1. **128**

$\text{Total} = 23 \times 4 + 36$
$= 92 + 36$
$= 128$
$\therefore$ Lucas had 128 stickers.

2. **$51 000**
Total attendance = 247 + 353
= 600
Total amount = $85 × 600
= $51 000
∴ total amount paid was $51 000.

3. **$355**
Cost of handbag = $213
Cost of dress = $213 ÷ 3
= $71
Total paid = $71 × 2 + $213
= $142 + $213
= $355
∴ Catherine will pay $355.

4. **20**
Adult admission = $12 × 30
= $360
Total cost of children = $520 − $360
= $160
No. of children = 160 ÷ 8
= 20
∴ 20 children attended.

5. **360 g**
Amount used = 1250 − 290
= 960
Amount per cake = 960 ÷ 8
= 120
∴ 120 grams per cake
∴ 360 grams for 3 cakes

6. **$216**
Difference in amounts = $24 − $20
= $4
Number of days = 36 ÷ 4
= 9
Total amount = $24 × 9
= $216
∴ they were originally given $216 each.

Key Skill  10 Lowest common multiple (pages 30–31)

1. **6**
You need the lowest common multiple of 2 and 3:
2: 2, 4, **6**, 8, …
3: 3, **6**, 9, …
∴ the smallest possible number is 6.

2. **15 seconds**
You need the lowest common multiple of 3 and 5:
3: 3, 6, 9, 12, **15**, 18, …
5: 5, 10, **15**, …
∴ the next time they blink is at 15 seconds.

3. **21 days**
You need the lowest common multiple of 3 and 7:
3: 3, 6, 9, 12, 15, 18, **21**, 24, …
7: 7, 14, **21**, 28, …
∴ they will ride again after 21 days.

4. **24**
You need the lowest common multiple of 3 and 8:
3: 3, 6, 9, 12, 15, 18, 21, **24**, …
8: 8, 16, **24**, 32, …
∴ the smallest number of bananas is 24.

5. **40**
You need the lowest common multiple of 8 and 10:
8: 8, 16, 24, 32, **40**, 48, …
10: 10, 20, 30, **40**, 50, …
∴ the smallest number of golf balls is 40.

6. **12**
You need the lowest common multiple of 3, 4 and 6:
3: 3, 6, 9, **12**, 15, …
4: 4, 8, **12**, 16, …
6: 6, **12**, 18, …
∴ the next time will be in 12 nights.

Key Skill  11 Greatest common factor (pages 32–33)

1. **12**
You need the greatest common factor of 48 and 36:
48: 1, 2, 3, 4, 6, 8, **12**, 16, 24, 48
36: 1, 2, 3, 4, 6, 9, **12**, 18, 36
∴ the largest number of groups is 12
(which will have 4 boys and 3 girls).

2. **12**
You need the greatest common factor of 24 and 36:
24: 1, 2, 3, 4, 6, 8, **12**, 16, 24
36: 1, 2, 3, 4, 6, 9, **12**, 18, 36
∴ the largest number of bouquets is 12.

3. **10**
You need the greatest common factor of 30 and 20:
30: 1, 2, 3, 5, 6, **10**, 15, 30
20: 1, 2, 3, 4, 5, **10**, 20
∴ the largest number of bathrooms is 10.

4. **20**
You need the greatest common factor of 60 and 80:
60: 1, 2, 3, 4, 5, 6, 10, 12, 15, **20**, 30, 60
80: 1, 2, 4, 5, 8, 10, 16, **20**, 40, 80
∴ the largest number of packs is 20.

5. **24**
You need the greatest common factor of 48 and 72:
48: 1, 2, 3, 4, 6, 8, 12, 16, **24**, 48
72: 1, 2, 3, 4, 6, 8, 9, 12, 18, **24**, 36, 72
∴ the largest number of people is 24.

6. **6**
You need the greatest common factor of 30, 18 and 12:
30: 1, 2, 3, 5, **6**, 10, 15, 30
18: 1, 2, 3, **6**, 9, 18
12: 1, 2, 3, 4, **6**, 12
$\therefore$ the largest number of arrangements is 6.

Key Skill 12 Unitary method A

(pages 34–35)

1. **$70**
Cost of 8 T-shirts = $56
Cost of 1 T-shirt = $56 ÷ 8
= $7
Cost of 10 T-shirts = $7 × 10
= $70
$\therefore$ the cost is $70.

2. **54 tonnes**
Mass of 4 trucks = 36
Mass of 1 truck = 36 ÷ 4
= 9
Mass of 6 trucks = 9 × 6
= 54
$\therefore$ the mass is 54 tonnes.

3. **5 cm**
Length in 8 weeks = 4
Length in 1 week = 4 ÷ 8
$= \frac{1}{2}$
Length in 10 weeks $= \frac{1}{2} \times 10$
= 5
$\therefore$ it would grow 5 cm.

4. **$22.50**
Cost of 10 hamburgers = $37.50
Cost of 1 hamburger = $37.50 ÷ 10
= $3.75
Cost of 6 hamburgers = $3.75 × 6
= $22.50
$\therefore$ the cost is $22.50.

5. **$12**
Cost of 3 tennis balls = $7.20
Cost of 1 tennis ball = $7.20 ÷ 3
= $2.40
Cost of 5 tennis balls = $2.40 × 5
= $12
$\therefore$ the cost is $12.

6. **15 minutes**
Three pieces means 2 cuts; six pieces means 5 cuts.
Time for 2 cuts = 6
Time for 1 cut = 6 ÷ 2
= 3
Time for 5 cuts = 5 × 3
= 15
$\therefore$ Larry takes 15 minutes.

Key Skill 13 Unitary method B

(pages 36–37)

1. **45 minutes**
Minutes for 6 men = 30
Minutes for 1 man = 30 × 6
= 180
Minutes for 4 men = 180 ÷ 4
= 45
$\therefore$ it would take 45 minutes.

2. **4 days**
Days for 12 soldiers = 5
Days for 1 soldier = 5 × 12
= 60
Days for 15 soldiers = 60 ÷ 15
= 4
$\therefore$ there is enough food for 4 days.

3. **$5**
Cost for 40 seniors = $4
Cost for 1 senior = $4 × 40
= $160
Cost for 32 seniors = $160 ÷ 32
= $40 ÷ 8
= $5
$\therefore$ it will cost $5 each.

4. **5 days**
Days for 20 workers = 6
Days for 1 worker = 6 × 20
= 120
Days for 24 workers = 120 ÷ 24
= 10 ÷ 2
= 5
$\therefore$ it will take 5 days.

5. **$60**
Cost for 4 occupants = $90
Cost for 1 occupant = $90 × 4
= $360
Cost for 6 occupants = $360 ÷ 6
= $60
$\therefore$ the cost will be $60 per person.

6. **4 days**
Days for 8 cockatiels = 5
Days for 1 cockatiel = 5 × 8
= 40
Days for 10 cockatiels = 40 ÷ 10
= 4
$\therefore$ the seed lasts for 4 days.

Key Skill 14 Unitary method C

(pages 38–39)

1. **480 m**
Metres by 3 fire-fighters in 2 days = 240
Metres by 1 fire-fighter in 2 days = 240 ÷ 3
= 80
Metres by 1 fire-fighter in 1 day = 80 ÷ 2
= 40
Metres by 4 fire-fighters in 1 day = 40 × 4
= 160
Metres by 4 fire-fighters in 3 days = 160 × 3
= 480
∴ they would clear 480 metres.

2. **60**
Ants to eat 1 pie in 2 hours = 40
Ants to eat 1 pie in 1 hour = 40 × 2
= 80
Ants to eat 3 pies in 1 hour = 80 × 3
= 240
Ants to eat 3 pies in 4 hours = 240 ÷ 4
= 60
∴ 60 ants would eat 3 pies in 4 hours.

3. **18 days**
Days for 6 workers to make 1 km = 4
Days for 6 workers to make 3 km = 4 × 3
= 12
Days for 1 worker to make 3 km = 12 × 6
= 72
Days for 4 workers to make 3 km = 72 ÷ 4
= 18
∴ it would take 18 days.

4. **20**
Eggs laid by 3 hens in 4 days = 6
Eggs laid by 1 hen in 4 days = 6 ÷ 3
= 2
Eggs laid by 1 hen in 8 days = 2 × 2
= 4
Eggs laid by 5 hens in 8 days = 4 × 5
= 20
∴ they would lay 20 eggs.

5. **4**
No. of men mow 4 greens in 20 min = 8
No. of men mow 1 green in 20 min = 8 ÷ 4
= 2
No. of men mow 6 greens in 20 min = 2 × 6
= 12
No. of men mow 6 greens in 60 min = 12 ÷ 3
= 4
∴ it would take 4 men.

6. **15 minutes**
Change 4 h to 240 min.
Minutes for 80% through 1 hole = 240
Minutes for 10% through 1 hole = 240 ÷ 8
= 30
Minutes for 10% through 2 holes = 30 ÷ 2
= 15
∴ it would take 15 minutes.

Revision Test 3
Level of difficulty—Average

(page 40)

1. **12 seconds**
Multiples of 3: 3, 6, 9, **12**, 15, …
Multiples of 4: 4, 8, **12**, 16, …
Multiples of 6: 6, **12**, 18, …
∴ they would ring together after 12 seconds.

2. **6**
You need the greatest common factor of 48 and 18:
48: 1, 2, 3, 4, **6**, 8, 12, 16, 24, 48
18: 1, 2, 3, **6**, 9, 18
∴ the largest number of gift packs is 6.

3. **180**
Cups from 8 kg coffee = 240
Cups from 1 kg coffee = 240 ÷ 8
= 30
Cups from 6 kg of coffee = 30 × 6
= 180
∴ Bobby can make 180 cups of coffee from 6 kg.

4. **36 minutes**
Minutes for 100 m^2 = 10
Minutes for 10 m^2 = 10 ÷ 10
= 1
Minutes for 360 m^2 = 1 × 36
= 36
∴ it will take 36 minutes.

5. **16 days**
Days for 8 men = 6
Days for 1 man = 6 × 8
= 48
Days for 3 men = 48 ÷ 3
= 16
∴ it would take 16 days.

6. **84**
Cans for 2 dogs for 1 day = 3
Cans for 2 dogs for 7 days = 3 × 7
= 21
Cans for 8 dogs for 7 days = 21 × 4
= 84
∴ it would take 84 cans.

Revision Test 4
Level of difficulty—Challenging (page 41)

1. **31**
Multiples of 2 plus 1: 3, 5, 7, 9, 11, 13, 15, 17, 19, 21, 23, 25, 27, 29, **31**, …
Multiples of 3 plus 1: 4, 7, 10, 13, 16, 19, 22, 25, 28, **31**, …
Multiples of 5 plus 1: 6, 11, 16, 21, 26, **31**, …
The first common number is 31.
∴ the smallest group is 31 students.

2. **36**
People for 10 days = 24
People for 1 day = 24 × 10
= 240
People for 4 days = 240 ÷ 4
= 60
Extra people = 60 − 24
= 36
∴ need to employ another 36 people.

3. **26**
Multiples of 3 plus 2: 5, 8, 11, 14, 17, 20, 23, **26**, 29, 32, …
Multiples of 5 plus 1: 6, 11, 16, 21, **26**, 31, …
Multiples of 2: 2, 4, 6, 8, 10, 12, 14, 16, 18, 20, 22, 24, **26**, …
The first common number is 26.
∴ the smallest number is 26.

4. **12 days**
Days for 4 machines on 2400 ha = 12
Days for 1 machine on 2400 ha = 12 × 4
= 48
Days for 6 machines on 2400 ha = 48 ÷ 6
= 8
Days for 6 machines on 1200 ha = 8 ÷ 2
= 4
Days for 6 machines on 3600 ha = 4 × 3
= 12
∴ it will take 12 days.

5. **12**
You need the greatest common factor of 8, 16 and 12:
Factors of 8: 1, 2, **4**, 8
Factors of 16: 1, 2, **4**, 8, 16
Factors of 12: 1, 2, 3, **4**, 6, 12
The three varieties can be divided into four packs.
Total number of muffins = 4 × 12
= 48
Number of blueberry = 48 − (8 + 16 + 12)
= 48 − 36
= 12
∴ he bakes 12 blueberry muffins.
(each pack will have 2 chocolate, 4 strawberry, 3 banana and 3 blueberry.)

6. **19th**
Multiples of 1 plus 1: 2, 3, 4, 5, 6, **7**, 8, 9, 10, 11, 12, **13**, 14, 15, 16, 17, 18, **19**, …
Multiples of 2 plus 1: 3, 5, **7**, 9, 11, **13**, 15, 17, **19**, 21, 23, 25, 27, …
Multiples of 3 and 1: 4, **7**, 10, **13**, 16, **19** …
They all step on the 1st, 7th, 13th, 19th step.
∴ the fourth step altogether is the 19th step.

Key Skill 15 Fractions: Addition
(pages 42–43)

1. $\mathbf{1\frac{11}{20}}$ **h**
$$\text{Total time} = \frac{3}{4} + \frac{4}{5} = \frac{15 + 16}{20} = \frac{31}{20} = 1\frac{11}{20}$$
∴ she studied for $1\frac{11}{20}$ h.

2. $\mathbf{5\frac{3}{4}}$ **h**
$$\text{Total time} = 3\frac{1}{4} + 2\frac{1}{2} = 5 + \frac{1 + 2}{4} = 5\frac{3}{4}$$
∴ the car was worked on for $5\frac{3}{4}$ h.

3. $\mathbf{7\frac{7}{10}}$ **km**
$$\text{Total distance} = 2\frac{1}{5} + 5\frac{1}{2} = 7 + \frac{2 + 5}{10} = 7\frac{7}{10}$$
∴ the total distance was $7\frac{7}{10}$ km.

4. $\mathbf{5\frac{1}{6}}$ **cups**
$$\text{Total quantity} = 1\frac{1}{2} + 3\frac{2}{3} = 4 + \frac{3 + 4}{6} = 4 + \frac{7}{6} = 4 + 1\frac{1}{6} = 5\frac{1}{6}$$
∴ she needs $5\frac{1}{6}$ cups.

5. $\mathbf{\frac{11}{12}}$ **of a pizza**

$$\text{Total amount} = \frac{1}{4} + \frac{2}{3} = \frac{3 + 8}{12} = \frac{11}{12}$$

∴ he ate $\frac{11}{12}$ of a pizza.

6. $\mathbf{10\frac{1}{12}}$

$$\text{Total number} = 3\frac{1}{2} + 4\frac{1}{3} + 2\frac{1}{4} = 9 + \frac{6 + 4 + 3}{12} = 9 + \frac{13}{12} = 9 + 1\frac{1}{12} = 10\frac{1}{12}$$

∴ he consumed $10\frac{1}{12}$ bottles.

Key Skill 16 Fractions: Subtraction (pages 44–45)

1. $\mathbf{1\frac{1}{4}}$ **km**

$$\text{Distance} = 2 - \frac{3}{4} = 1\frac{1}{4}$$

∴ Larry can swim $1\frac{1}{4}$ km further.

2. $\mathbf{5\frac{1}{4}}$ **m**

$$\text{Length} = 10 - 4\frac{3}{4} = 6 - \frac{3}{4} = 5\frac{1}{4} \text{ metres}$$

∴ there are $5\frac{1}{4}$ metres left on the roll.

3. $\mathbf{3\frac{1}{4}}$ **L**

$$\text{Amount} = 9\frac{1}{2} - 6\frac{1}{4} = 3 + \frac{1}{2} - \frac{1}{4} = 3 + \frac{2 - 1}{4} = 3 + \frac{1}{4} = 3\frac{1}{4}$$

∴ Terry got $3\frac{1}{4}$ litres of milk from the goat.

4. $\mathbf{7\frac{1}{4}}$ **bags**

$$\text{Amount} = 12\frac{3}{4} - 5\frac{1}{2} = 7 + \frac{3}{4} - \frac{1}{2} = 7 + \frac{3 - 2}{4} = 7 + \frac{1}{4} = 7\frac{1}{4}$$

∴ $7\frac{1}{4}$ more bags of white flour were used.

5. $\mathbf{3\frac{1}{4}}$ **hours**

$$\text{Time} = 9\frac{3}{4} - 6\frac{1}{2} = 3 + \frac{3}{4} - \frac{1}{2} = 3 + \frac{3 - 2}{4} = 3 + \frac{1}{4} = 3\frac{1}{4}$$

∴ Owen took $3\frac{1}{4}$ hours longer.

6. $2\frac{1}{4}$ **minutes**

$\text{Time} = 10\frac{1}{2} - 8\frac{1}{4}$

$= 2 + \frac{1}{2} - \frac{1}{4}$

$= 2 + \frac{2-1}{4}$

$= 2 + \frac{1}{4}$

$= 2\frac{1}{4}$

$\therefore$ Ken waited $2\frac{1}{4}$ minutes.

Key Skill  17 Fractions: Multiplication A

(pages 46–47)

1. 54

$\text{Number of blue marbles} = \frac{3}{4} \times 72$

$= \frac{3}{_1\cancel{4}} \times \frac{\cancel{72}^{18}}{1}$

$= 54$

$\therefore$ there were 54 blue marbles.

2. 9

$\text{Fraction not on excursion} = 1 - \frac{2}{3}$

$= \frac{1}{3}$

$\text{Students not on excursion} = \frac{1}{3} \times 27$

$= \frac{1}{_1\cancel{3}} \times \frac{\cancel{27}^{9}}{1}$

$= 9$

$\therefore$ 9 students are not going on the excursion.

3. 24

$\text{Number of cola cans} = \frac{2}{5} \times 60$

$= \frac{2}{_1\cancel{5}} \times \frac{\cancel{60}^{12}}{1}$

$= 24$

$\therefore$ Leonard bought 24 cola cans.

4. 90

$\text{Fraction who do not have sister} = 1 - \frac{3}{8}$

$= \frac{5}{8}$

$\text{Students with no sister} = \frac{5}{8} \times 144$

$= \frac{5}{_1\cancel{8}} \times \frac{\cancel{144}^{18}}{1}$

$= 90$

$\therefore$ 90 students do not have a sister.

5. \$140

$\text{Fraction to other charities} = 1 - \frac{5}{12}$

$= \frac{7}{12}$

$\text{Amount to other charities} = \frac{7}{12} \times 240$

$= \frac{7}{_1\cancel{12}} \times \frac{\cancel{240}^{20}}{1}$

$= 140$

$\therefore$ \$140 will go to the other charities.

6. 190

$\text{Two out of three} = \frac{2}{3}$

$\text{Fraction of male members} = 1 - \frac{2}{3}$

$= \frac{1}{3}$

$\text{No. of male members} = \frac{1}{3} \times 570$

$= \frac{1}{_1\cancel{3}} \times \frac{\cancel{570}^{190}}{1}$

$= 190$

$\therefore$ there are 190 male members.

Key Skill  18 Fractions: Multiplication B

(pages 47–48)

1. $\frac{1}{6}$

$\text{Fraction for photocopier} = \frac{1}{4} \times \frac{2}{3}$

$= \frac{1}{_2\cancel{4}} \times \frac{\cancel{2}^{1}}{3}$

$= \frac{1}{6}$

$\therefore \frac{1}{6}$ of the money was used for the photocopier.

2. $\mathbf{4\frac{1}{2}}$ **m**

$$\begin{aligned}\text{Fraction not used} &= 1 - \frac{4}{7}\\ &= \frac{3}{7}\\ \text{Length} &= \frac{3}{7} \times 10\frac{1}{2}\\ &= \frac{3}{{}_1\cancel{7}} \times \frac{\cancel{21}^3}{2}\\ &= \frac{9}{2}\\ &= 4\frac{1}{2}\end{aligned}$$

$\therefore 4\frac{1}{2}$ metres of string was not used.

3. $\mathbf{\frac{1}{4}}$

$$\begin{aligned}\text{Fraction strawberry and coconut} &= \frac{1}{3} \times \frac{3}{4}\\ &= \frac{1}{{}_1\cancel{3}} \times \frac{\cancel{3}^1}{4}\\ &= \frac{1}{4}\end{aligned}$$

$\therefore \frac{1}{4}$ of the cupcakes had strawberry icing and coconut sprinkles.

4. $\mathbf{2\frac{1}{5}}$ **hours**

$$\begin{aligned}\text{Winner's time} &= \frac{3}{5} \times 3\frac{2}{3}\\ &= \frac{3}{5} \times \frac{11}{3}\\ &= \frac{{}^1\cancel{3}}{5} \times \frac{11}{\cancel{3}_1}\\ &= \frac{11}{5}\\ &= 2\frac{1}{5}\end{aligned}$$

$\therefore$ the winner took $2\frac{1}{5}$ hours.

5. $\mathbf{\frac{1}{2}}$

$$\begin{aligned}\text{Fraction dog and cat owner} &= \frac{3}{4} \times \frac{2}{3}\\ &= \frac{{}^1\cancel{3}}{{}_2\cancel{4}} \times \frac{\cancel{2}^1}{\cancel{3}_1}\\ &= \frac{1}{2}\end{aligned}$$

$\therefore \frac{1}{2}$ of the children have both a dog and cat.

6. $\mathbf{\frac{1}{4}}$

$$\begin{aligned}\text{Fraction with blue eyes} &= 1 - \frac{1}{4}\\ &= \frac{3}{4}\\ \text{Fraction blonde blue-eyed} &= \frac{3}{4} \times \frac{1}{3}\\ &= \frac{{}^1\cancel{3}}{4} \times \frac{1}{\cancel{3}_1}\\ &= \frac{1}{4}\end{aligned}$$

$\therefore \frac{1}{4}$ of the students have blonde hair and blue eyes.

Key Skill 19 Fractions: Division

(pages 50–51)

1. $\mathbf{\frac{1}{2}}$ km

$$\begin{aligned}\text{Distance each morning} &= 2\frac{1}{2} \div 5\\ &= \frac{5}{2} \div \frac{5}{1}\\ &= \frac{{}^1\cancel{5}}{2} \times \frac{1}{\cancel{5}_1}\\ &= \frac{1}{2}\end{aligned}$$

$\therefore$ Grace swam $\frac{1}{2}$ km each morning.

2. 27 days

$$\begin{aligned}\text{Number of days} &= 4\frac{1}{2} \div \frac{1}{6}\\ &= \frac{9}{2} \times \frac{6}{1}\\ &= \frac{9}{{}_1\cancel{2}} \times \frac{\cancel{6}^3}{1}\\ &= 27\end{aligned}$$

$\therefore$ the food lasts 27 days.

3. 10

$$\begin{aligned}\text{Number of loads} &= 2 \div \frac{1}{5}\\ &= \frac{2}{1} \times \frac{5}{1}\\ &= 10\end{aligned}$$

$\therefore$ the powder will last 10 loads.

4. **32**

$$\text{Number of glasses} = 12 \div \frac{3}{8}$$
$$= \frac{12}{1} \times \frac{8}{3}$$
$$= \frac{^{4}\cancel{12}}{1} \times \frac{8}{\cancel{3}_{1}}$$
$$= 32$$

∴ Chen can fill 32 glasses.

5. **6**

$$\text{Number of batches} = 10 \div 1\frac{2}{3}$$
$$= 10 \div \frac{5}{3}$$
$$= \frac{10}{1} \times \frac{3}{5}$$
$$= \frac{^{2}\cancel{10}}{1} \times \frac{3}{\cancel{5}_{1}}$$
$$= 6$$

∴ Jack can make 6 batches.

6. **4 weeks**

$$\text{Days} = 70 \div 3\frac{1}{2}$$
$$= 70 \div \frac{7}{2}$$
$$= \frac{70}{1} \times \frac{2}{7}$$
$$= \frac{^{10}\cancel{70}}{1} \times \frac{2}{\cancel{7}_{1}}$$
$$= 20$$

∴ as there are 5 weekdays each week, then Jose walked for 4 weeks.

Key Skill 20 Fractions: Addition and subtraction

(pages 52–53)

1. $\mathbf{\frac{3}{10}}$

$$\text{Fraction} = 1 - (\frac{1}{2} + \frac{1}{5})$$
$$= 1 - \frac{5 + 2}{10}$$
$$= 1 - \frac{7}{10}$$
$$= \frac{3}{10}$$

∴ Craig kicked $\frac{3}{10}$ of the goals.

2. $\mathbf{\frac{3}{10}}$

$$\text{Fraction} = 1 - (\frac{3}{5} + \frac{1}{10})$$
$$= 1 - \frac{6 + 1}{10}$$
$$= 1 - \frac{7}{10}$$
$$= \frac{3}{10}$$

∴ the team lost $\frac{3}{10}$ of the games.

3. $\mathbf{\frac{3}{20}}$

$$\text{Fraction} = 1 - (\frac{1}{4} + \frac{3}{5})$$
$$= 1 - \frac{5 + 12}{20}$$
$$= 1 - \frac{17}{20}$$
$$= \frac{3}{20}$$

∴ $\frac{3}{20}$ of all accidents.

4. $\mathbf{1\frac{1}{4}}$

$$\text{Fraction} = 5 - (1\frac{1}{2} + 2\frac{1}{4})$$
$$= 5 - (3 + \frac{2 + 1}{4})$$
$$= 5 - 3\frac{3}{4}$$
$$= 2 - \frac{3}{4}$$
$$= 1\frac{1}{4}$$

∴ $1\frac{1}{4}$ metres of material remains.

5. $\mathbf{\frac{1}{4}}$

$$\text{Fraction musks} = 1 - (\frac{1}{3} + \frac{1}{4} + \frac{1}{6})$$
$$= 1 - \frac{4 + 3 + 2}{12}$$
$$= 1 - \frac{9}{12}$$
$$= \frac{3}{12}$$
$$= \frac{1}{4}$$

∴ $\frac{1}{4}$ of the lollies are musks.

6. $\frac{23}{24}$

$$\text{Fraction} = 3 - (\frac{3}{4} + \frac{2}{3} + \frac{5}{8})$$
$$= 3 - \frac{18 + 16 + 15}{24}$$
$$= 3 - \frac{49}{24}$$
$$= 3 - 2\frac{1}{24}$$
$$= 1 - \frac{1}{24}$$
$$= \frac{23}{24}$$

$\therefore \frac{23}{24}$ of a pizza was not eaten.

Key Skill 21 Fractions: Addition, subtraction and multiplication

(pages 54–55)

1. **2 m**

$$\text{Fraction above water} = 1 - (\frac{1}{6} + \frac{1}{2})$$
$$= 1 - \frac{1 + 3}{6}$$
$$= 1 - \frac{4}{6}$$
$$= \frac{2}{6}$$
$$= \frac{1}{3}$$
$$\text{Length} = \frac{1}{3} \times 6$$
$$= 2$$

$\therefore$ the pole is 2 metres above the water level.

2. **30**

$$\text{Fraction walkers} = 1 - (\frac{5}{8} + \frac{1}{3})$$
$$= 1 - \frac{15 + 8}{24}$$
$$= 1 - \frac{23}{24}$$
$$= \frac{1}{24}$$
$$\text{Number of walkers} = \frac{1}{24} \times 720$$
$$= \frac{1}{{}_{1}\cancel{2}\cancel{24}} \times \frac{\cancel{720}^{\cancel{60}\,30}}{1}$$
$$= 30$$

$\therefore$ 30 students walk to school.

3. **20**

$$\text{Fraction green} = 1 - (\frac{1}{3} + \frac{1}{4})$$
$$= 1 - \frac{4 + 3}{12}$$
$$= 1 - \frac{7}{12}$$
$$= \frac{5}{12}$$
$$\text{Green} = \frac{5}{12} \times 48$$
$$= \frac{5}{{}_{1}\cancel{12}} \times \frac{\cancel{48}^{4}}{1}$$
$$= 20$$

$\therefore$ there are 20 green balls.

4. **240**

$$\text{Fraction mandarine} = 1 - (\frac{1}{3} + \frac{2}{5})$$
$$= 1 - \frac{5 + 6}{15}$$
$$= 1 - \frac{11}{15}$$
$$= \frac{4}{15}$$
$$\text{Mandarine trees} = \frac{4}{15} \times 900$$
$$= \frac{4}{{}_{1}\cancel{15}} \times \frac{\cancel{900}^{60}}{1}$$
$$= 240$$

$\therefore$ there are 240 mandarine trees.

5. **$420**

$$\text{Fraction remaining} = 1 - (\frac{1}{4} + \frac{2}{5})$$
$$= 1 - \frac{5 + 8}{20}$$
$$= 1 - \frac{13}{20}$$
$$= \frac{7}{20}$$

$$\text{Amount remaining} = \frac{7}{20} \times 1200$$
$$= \frac{7}{{}_1\cancel{20}} \times \frac{\cancel{1200}^{60}}{1}$$
$$= 420$$
∴ Hugo has \$420 remaining.

6. **75**

'three in eight' $= \frac{3}{8}$ and 'three in five' $= \frac{3}{5}$

$$\text{Fraction children} = 1 - (\frac{3}{8} + \frac{3}{5})$$
$$= 1 - \frac{15 + 24}{40}$$
$$= 1 - \frac{39}{40}$$
$$= \frac{1}{40}$$
$$\text{Children} = \frac{1}{40} \times 3000$$
$$= \frac{1}{{}_1\cancel{40}} \times \frac{\cancel{3000}^{75}}{1}$$
$$= 75$$
∴ there are 75 children on the cruise.

Key Skill  22 Fractions using unitary method

(pages 56–57)

1. **400**

$$\text{Fraction still to read} = 1 - \frac{3}{8}$$
$$= \frac{5}{8}$$
Five-eighths of the book = 250
One-eighth of the book = 250 ÷ 5
= 50
Eight-eighths of the book = 50 × 8
= 400
∴ there are 400 pages in the book.

2. **2700 L**

$$\text{Fraction yet to fill} = 1 - \frac{3}{5}$$
$$= \frac{2}{5}$$
Two-fifths of the tank = 1800
One-fifth of the tank = 1800 ÷ 2
= 900
Three-fifths of the tank = 900 × 3
= 2700
∴ there was 2700 L before it rained.

3. **250 km**

$$\text{Fraction yet to travel} = 1 - \frac{1}{5}$$
$$= \frac{4}{5}$$
Four-fifths of distance = 200
One-fifh of distance = 200 ÷ 4
= 50
Five-fifths of distance = 50 × 5
= 250
∴ the total distance is 250 km.

4. **480**

$$\text{Fraction females} = 1 - \frac{4}{5}$$
$$= \frac{1}{5}$$
One-fifth of audience = 120
Four-fifths of audience = 120 × 4
= 480
∴ there were 480 males.

5. **\$210**

$$\text{Fraction yet to spend} = 1 - \frac{4}{7}$$
$$= \frac{3}{7}$$
Three-sevenths of money = \$90
One-seventh of money = \$90 ÷ 3
= \$30
Seven-sevenths of money = \$30 × 7
= \$210
∴ Sofia was given \$210.

6. **32**

$$\text{Fraction who can swim} = 1 - \frac{3}{8}$$
$$= \frac{5}{8}$$
$$\text{Difference} = \frac{5}{8} - \frac{3}{8}$$
$$= \frac{2}{8}$$
$$= \frac{1}{4}$$
One-quarter of students = 8
Four-quarters of students = 8 × 4
= 32
∴ there are 32 students in the group.

Key Skill 23 Ratio: Dividing quantities in a given ratio (pages 58–59)

1. **18**

$$\text{Total parts} = 5 + 3 = 8$$

$$\text{Number of females} = \frac{3}{\cancel{8}_1} \times \frac{\cancel{48}^6}{1} = 18$$

∴ there are 18 female students.

2. **72**

$$\text{Ratio of scones} = 2:6 = 1:3$$

∴ Kevin : Phillip is 1:3

$$\text{Total parts} = 1 + 3 = 4$$

$$\text{Phillip's scones} = \frac{3}{\cancel{4}_1} \times \frac{\cancel{96}^{24}}{1} = 72$$

∴ Phillip made 72 scones.

3. **2 L**

$$\text{Total parts} = 1 + 4 = 5$$

$$\text{Water} = \frac{4}{5} \text{ of } 2\frac{1}{2} = \frac{{}^2\cancel{4}}{{}_1\cancel{5}} \times \frac{\cancel{5}^1}{\cancel{2}_1} = 2$$

∴ Lee used 2 litres of water.

4. **$20 000**

$$\text{Ratio of contributions} = \$8:\$2 = 4:1$$

∴ Jason : Natasha is 4 : 1

$$\text{Total parts} = 4 + 1 = 5$$

$$\text{Jason's winnings} = \frac{4}{\cancel{5}_1} \times \frac{\cancel{25000}^{5000}}{1} = 20\,000$$

∴ Jason will receive $20 000.

5. **15 cm**

Perimeter is 66 which means sum of length and width is 33 cm.

$$\text{Total parts} = 6 + 5 = 11$$

$$\text{Width} = \frac{5}{\cancel{11}_1} \times \frac{\cancel{33}^3}{1} = 15$$

∴ the width is 15 cm.

6. **1560**

$$\text{Ratio of rolls} = 72:36:6 = 12:6:1$$

∴ white : multigrain : wholemeal = 12 : 6 : 1

$$\text{Total parts} = 12 + 6 + 1 = 19$$

$$\text{Number not multigrain} = 19 - 6 = 13$$

∴ 13 out of 19 not multigrain

$$\text{Not multigrain} = \frac{13}{\cancel{19}_1} \times \frac{\cancel{2200}^{120}}{1} = 1560$$

∴ 1560 rolls were not multigrain.

Revision Test 5
Level of difficulty—Average (page 60)

1. $\mathbf{7\frac{19}{20}}$ **km**

$$\text{Total distance} = 4\frac{1}{5} + 3\frac{3}{4} = 7 + \frac{1}{5} + \frac{3}{4} = 7 + \frac{4 + 15}{20} = 7 + \frac{19}{20} = 7\frac{19}{20}$$

∴ Rebecca jogged $7\frac{19}{20}$ km.

2. $\mathbf{\frac{7}{8}}$ **m**

$$\text{Fraction remaining} = 1 - \frac{3}{4} = \frac{1}{4}$$

$$\text{Length remaining} = \frac{1}{4} \times 3\frac{1}{2} = \frac{1}{4} \times \frac{7}{2} = \frac{7}{8}$$

∴ $\frac{7}{8}$ m remained.

3. $\frac{13}{60}$

Fraction bank owned $= 1 - (\frac{1}{4} + \frac{1}{5} + \frac{1}{3})$

$= 1 - \frac{15 + 12 + 20}{60}$

$= 1 - \frac{47}{60}$

$= \frac{13}{60}$

$\therefore$ the bank owns $\frac{13}{60}$.

4. **48 L**

Litres in three-quarters of tank = 36
Litres in one-quarter of tank = 36 ÷ 3
= 12
Litres in four-quarters of tank = 12 × 4
= 48

$\therefore$ the tank holds 48 litres.

5. **6 L**

Litres in half a tank = 18
Litres in full tank = 18 × 2
= 36

Litres in two-thirds of a tank $= \frac{2}{_1\cancel{3}} \times \frac{\cancel{36}^{12}}{1}$

= 24

$\therefore$ Litres in two-thirds tank is 24 L.

Difference between $\frac{1}{2}$ and $\frac{2}{3}$ tank = 24 − 18
= 6

Oil added to drum is 6 L.

6. **24**

Total parts = 3 + 4
= 7

Number of girls $= \frac{4}{7}$ of 42

$= \frac{4}{_1\cancel{7}} \times \frac{\cancel{42}^{6}}{1}$

= 24

$\therefore$ there were 24 girls.

Revision Test 6
Level of difficulty—Challenging (page 61)

1. **3**

Cups of cream for 1 person $= 1\frac{3}{4} \div 7 = \frac{7}{4} \div \frac{7}{1}$

$= \frac{^1\cancel{7}}{4} \times \frac{1}{\cancel{7}^1}$

$= \frac{1}{4}$

Cups of cream for 12 people $= \frac{1}{4} \times 12$

$= \frac{1}{_1\cancel{4}} \times \frac{\cancel{12}^3}{1}$

= 3

$\therefore$ 3 cups of cream are needed.

2. **$62.50**

Money on clothes $= \frac{2}{5} \times 2500$

$= \frac{2}{_1\cancel{5}} \times \frac{\cancel{2500}^{500}}{1}$

= 1000

Money on jeans $= \frac{1}{4} \times 1000$

= 250

Cost of each pair = 250 ÷ 4
= 62.5

$\therefore$ Taylor spent $62.50 on each pair.

3. **320**

Marbles in 4 bags = 24 × 4
= 96

$\therefore$ he has 96 marbles.

Two-fifths of remaining = 96
One-fifth of remaining = 96 ÷ 2
= 48
Five-fifths of remaining = 48 × 5
= 240

$\therefore$ he had 240 marbles after the game.

Three-quarters of original = 240
One-quarter of original = 240 ÷ 3
= 80
Four-quarters of original = 80 × 4
= 320

$\therefore$ he had 320 marbles in the beginning.

4. **32**

METHOD 1:

Mark's father ate $\frac{1}{2}$

Mark ate $\frac{1}{2}$ of remaining $\frac{1}{2} = \frac{1}{2} \times \frac{1}{2} = \frac{1}{4}$

Mark's mother ate $\frac{1}{2}$ of remaining $\frac{1}{4}$

$= \frac{1}{2} \times \frac{1}{4} = \frac{1}{8}$

Amount eaten so far $= \frac{1}{2} + \frac{1}{4} + \frac{1}{8} = \frac{7}{8}$

Remaining $= \frac{1}{8}$ = 4 pieces

Total chocolate = 8 × 4 = 32 pieces

METHOD 2:
Pieces eaten by Mark's sister = 4
∴ Mark's mother ate the same number of pieces (4) as his sister
Pieces before mother eats = 2 × 4
= 8
∴ Mark ate the same number of pieces (8) as remained after he had eaten.
Pieces before Mark eats = 2 × 8
= 16
∴ Mark ate the same number of pieces (16) as remained after he had eaten.
Pieces before Mark's father eats = 2 × 16
= 32
∴ there were 32 pieces in the chocolate block.

5. **12**
Suggested ratio =
Albert : Ben : Charles = 3 : 1 : 2
Total parts = 3 + 1 + 2
= 6
Number of Ben's stickers = $\frac{1}{6}$ of 72
= 12
∴ Ben's share was 12 stickers.

6. **45**
METHOD 1:
Fraction sold on first day = $\frac{1}{3}$
Fraction unsold on first day = $\frac{2}{3}$
Fraction sold on second day = $\frac{1}{3} \times \frac{2}{3}$
$= \frac{2}{9}$
Fraction sold on both days = $\frac{1}{3} + \frac{2}{9}$
$= \frac{5}{9}$
Fraction unsold on both days = $1 - \frac{5}{9}$
$= \frac{4}{9}$
$\frac{4}{9}$ of total lemons = 20
$\frac{1}{9}$ of total lemons = 5
$\frac{9}{9}$ of total lemons = 45
∴ Brodie had originally 45 lemons.
METHOD 2:
On the second day, Brodie sold one-third of his available lemons.
Therefore, two-thirds of 2nd-day lemons = 20
One-third of 2nd-day lemons = 10
Total of 2nd-day lemons = 30
This means he finished the first day with 30 lemons.
As he sold one-third of his lemons on first day, then he did not sell two-thirds.
Therefore, two-thirds of 1st-day lemons = 30
One-third of 1st-day lemons = 15
Total of 1st-day lemons = 45
Therefore, Brodie originally had 45 lemons for sale.

Key Skill  24 Decimals: Addition

(pages 62–63)

1. **98.1 kg**
Total Mass = 46.8 + 51.3
= 98.1

$$\begin{array}{r} 46.8 \\ +\ 51.3 \\ \hline 98.1 \end{array}$$

∴ the total mass is 98.1 kg.

2. **101.5 L**
Total litres = 32.7 + 38.9 + 29.9
= 101.5

$$\begin{array}{r} 32.7 \\ 38.9 \\ +\ 29.9 \\ \hline 101.5 \end{array}$$

∴ 101.5 L was purchased.

3. **$174.70**
Total amount = $69.95 + $44.90 + $59.85
= $174.70

$$\begin{array}{r} 69.95 \\ 44.90 \\ +\ 59.85 \\ \hline 174.70 \end{array}$$

∴ the total amount was $174.70.

4. **1.54 m**
Height = 1.31 + 0.08 + 0.15
= 1.54

$$\begin{array}{r} 1.31 \\ 0.08 \\ +\ 0.15 \\ \hline 1.54 \end{array}$$

∴ she was 1.54 metres.

5. **15.55 km**
Distance = 4.8 + 6.7 + 4.05
= 15.55

$$\begin{array}{r} 4.80 \\ 6.70 \\ +\ 4.05 \\ \hline 15.55 \end{array}$$

∴ the yacht sailed 15.55 km.

6. **$10.50**
Total money = $4.85 + $2.50 + $3.10 + $0.05

$$\begin{array}{r} 4.85 \\ 2.50 \\ 3.10 \\ +\ 0.05 \\ \hline 10.50 \end{array}$$

∴ Wen handed over $10.50.

Key Skill  Decimals: Subtraction

(pages 64–65)

1. **0.11 s**
   Difference = 0.32 − 0.21
   = 0.11
   $$\begin{array}{r} 0.32 \\ -\ 0.21 \\ \hline 0.11 \end{array}$$
   ∴ the difference was 0.11 seconds.

2. **147.7 km**
   Distance = 215 − 67.3
   = 147.7
   $$\begin{array}{r} 215.0 \\ -\ \ 67.3 \\ \hline 147.7 \end{array}$$
   ∴ Benny still has 147.7 km to travel.

3. **6.6°**
   Temperature rise = 18.3 − 11. 7
   = 6.6
   $$\begin{array}{r} 18.3 \\ -\ 11.7 \\ \hline 6.6 \end{array}$$
   ∴ the temperature rose 6.6 degrees.

4. **1.85 L**
   Remains = 2.6 − 0.75
   = 1.85
   $$\begin{array}{r} 2.60 \\ -\ 0.75 \\ \hline 1.85 \end{array}$$
   ∴ there is still 1.85 L in the jug.

5. **5.6 million**
   Difference = 13.9 mill − 8.3 mill
   = 5.6 mill
   $$\begin{array}{r} 13.9 \\ -\ \ 8.3 \\ \hline 5.6 \end{array}$$
   ∴ China made 5.6 million more cars.

6. **8.25 cm**
   Thickness = 9 − 0.3 − 0.45
   = 8.35
   $$\begin{array}{r} 9.0 \\ -\ 0.3 \\ \hline 8.7 \end{array} \qquad \begin{array}{r} 8.70 \\ -\ 0.45 \\ \hline 8.25 \end{array}$$
   ∴ it was 8.25 cm thick.

Key Skill  Decimals: Multiplication A

(pages 66–67)

1. **13.5 m**
   Distance = 2.7 × 5
   = 13.5
   $$\begin{array}{r} 27 \\ \times\ \ \ 5 \\ \hline 135 \end{array}$$
   ∴ the distance is 13.5 m.

2. **86 kg**
   Total mass = 2.15 × 40
   = 86.00
   $$\begin{array}{r} 215 \\ \times\ \ 40 \\ \hline 8600 \end{array}$$
   ∴ the pavers have a mass of 86 kg.

3. **$11.10**
   Total cost = $1.85 × 6
   = $11.10
   $$\begin{array}{r} 185 \\ \times\ \ \ 6 \\ \hline 1110 \end{array}$$
   ∴ the chocolates cost $11.10.

4. **15.2 m**
   Length = 1.9 × 8
   = 15.2
   $$\begin{array}{r} 19 \\ \times\ \ 8 \\ \hline 152 \end{array}$$
   ∴ the total length is 15.2 metres.

5. **22.6 g**
   Mass = 11.3 × 2
   = 22.6
   ∴ the mass of the sinker is 22.6 grams.

6. **90%**
   Percentage = 72 × 1.25
   = 90
   $$\begin{array}{r} 125 \\ \times\ \ \ 72 \\ \hline 250 \\ 8750 \\ \hline 9000 \end{array}$$
   ∴ Sandy's mark was 90%.

Key Skill  Decimals: Multiplication B

(pages 68–69)

1. **$5.52**
   Cost = 6.90 × 0.8
   = 5.520
   $$\begin{array}{r} 690 \\ \times\ \ \ 8 \\ \hline 5520 \end{array}$$
   ∴ the cost was $5.52.

2. **$31.15**
   Cost = 8.90 × 3.5
   = 31.150
   $$\begin{array}{r} 890 \\ \times\ \ \ 35 \\ \hline 4450 \\ 26700 \\ \hline 31150 \end{array}$$
   ∴ the cost was $31.15.

3. **$7.88**
   Cost = 19.70 × 0.4
   = 7.880
   $$\begin{array}{r} 1970 \\ \times\ \ \ \ 4 \\ \hline 7880 \end{array}$$
   ∴ the cost was $7.88.

4. **161.4 cm**
   Height = 134.5 × 1.2
   = 161.40
   $$\begin{array}{r} 1345 \\ \times\ \ \ 12 \\ \hline 16140 \end{array}$$
   ∴ Ingrid's height is 161.4 cm.

5. **9.1 L**
Total amount = 6.5 × 1.4
= 9.10

65
× 14
260
650
910

∴ 9.1 litres of oil used.

6. **$123.25**
Pay = 14.50 × 8.5
= 123.250

1450
× 85
7250
116000
123250

∴ he is paid $123.25.

Key Skill  Decimals: Division A

(pages 70–71)

1. **7.7 km**
Distance = 23.1 ÷ 3
= 7.7

$3\overline{)23.1}$ = 7.7

∴ the distance was 7.7 km.

2. **$85.70**
Amount = $428.50 ÷ 5
= $85.70

$5\overline{)428.50}$ = 85.70

∴ they each paid $85.70.

3. **$14.50**
Cost = $58 ÷ 4
= $14.50

$4\overline{)58.00}$ = 14.50

∴ each ticket cost $14.50.

4. **3.75 kg**
Amount = 11.25 ÷ 3
= 3.75

$3\overline{)11.25}$ = 3.75

∴ each garden receives 3.75 kg.

5. **8.7 kg**
Amount = 34.8 ÷ 4
= 8.7

$4\overline{)34.8}$ = 8.7

∴ each batch used 8.7 kg of chocolate chips.

6. **$3.81**
Amount = $76.20 ÷ 20
= $3.81

$20\overline{)76.20}$ = 3.81

∴ she paid $3.81 each day.

Key Skill  Decimals: Division B

(pages 72–73)

1. **3**
Number of lengths = 2.4 ÷ 0.8
= 24 ÷ 8
= 3

∴ 3 lengths can be cut.

2. **8**
Number of glasses = 2 ÷ 0.25
= 200 ÷ 25
= 40 ÷ 5
= 8

∴ Phillipa filled 8 glasses.

3. **125**
Number of strides = 100 ÷ 0.8
= 1000 ÷ 8
= 500 ÷ 4
= 250 ÷ 2
= 125

∴ Harry needs to take 125 strides.

4. **25 L**
Number of litres = 1 ÷ 0.04
= 100 ÷ 4
= 25

∴ 25 litres of milk are needed.

5. **50 miles**
Number of miles = 80 ÷ 1.6
= 800 ÷ 16
= 400 ÷ 8
= 50

∴ there are 50 miles.

6. **20**
Number of cattle = 15 ÷ 0.75
= 1500 ÷ 75
= 500 ÷ 25
= 20

∴ Jim will put 20 cattle into the pasture.

Key Skill  Decimals: Addition and subtraction

(pages 74–75)

1. **$1.50**
Change = $40.00 − ($19.90 + $18.60)
= $1.50

19.90
+ 18.60
38.50

40.00
− 38.50
1.50

∴ the change will be $1.50.

2. **18.9 m**
Length = 50 − (12.4 + 18.7)

12.4
+ 18.7
31.1

50.0
− 31.1
18.9

∴ 18.9 metres remains.

3. **$30.40**
Spent = $38.50 + $19.90 + $11.20

$38.50
19.90
+ 11.20
$69.60

Remainder = \$100 − \$69.60

$$\begin{array}{r} \$100.00 \\ -\ \ 69.60 \\ \hline \$30.40 \end{array}$$

∴ \$30.40 remains on the card.

4. **\$110.65**

Raised = \$120.50 + \$355.80 + \$413.05

$$\begin{array}{r} \$120.50 \\ 355.80 \\ +\ \ 413.05 \\ \hline \$889.35 \end{array}$$

Money needed = \$1000 − \$889.35

$$\begin{array}{r} \$1000.00 \\ -\ \ 889.35 \\ \hline \$110.65 \end{array}$$

∴ the school needs another \$110.65.

5. **0.72 gigabytes**

Space = 4.3 − (1.83 + 1.75)

$$\begin{array}{r} 1.83 \\ +\ 1.75 \\ \hline 3.58 \end{array} \qquad \begin{array}{r} 4.30 \\ -\ 3.58 \\ \hline 0.72 \end{array}$$

∴ there is 0.72 gigabytes available.

6. **\$556.60**

Deductions = \$265.50 + \$142.60 + \$75.30

$$\begin{array}{r} \$265.50 \\ 142.60 \\ +\ \ 75.30 \\ \hline \$483.40 \end{array}$$

Remainder = \$1040 − \$483.40

$$\begin{array}{r} \$1000.00 \\ -\ 483.40 \\ \hline \$556.60 \end{array}$$

∴ she has \$556.60 remaining.

Key Skill 31 Decimals: Addition, subtraction and multiplication

(page 76–77)

1. **1.1 km**

Total distance = 3.1 × 3 + 2.4 × 4
= 9.3 + 9.6
= 18.9

Remainder = 20 − 18.9

$$\begin{array}{r} 20.0 \\ -\ 18.9 \\ \hline 1.1 \end{array}$$

∴ Jasmin is 1.1 km short of her goal.

2. **\$71.60**

Total cost = \$12.90 × 10 + \$49.90 × 6
= \$129 + \$299.40
= \$428.40

Remainder = \$500 − \$428.40

$$\begin{array}{r} 500.00 \\ -\ 428.40 \\ \hline 71.60 \end{array}$$

∴ \$71.60 remains.

3. **\$18.50**

Paint and brushes = \$59.95 × 4 + \$11.90 × 3
= \$275.50

Cost of roller = \$294 − \$275.50

$$\begin{array}{r} 294.00 \\ -\ 275.50 \\ \hline 18.50 \end{array}$$

∴ the roller cost \$18.50.

4. **\$17.95**

Books and pens = \$1.25 × 4 + \$2.05 × 3
= \$11.15

Cost of calculator = \$29.10 − \$11.15

$$\begin{array}{r} 29.10 \\ -\ 11.15 \\ \hline 17.95 \end{array}$$

∴ the calculator cost \$17.95.

5. **\$2.85**

Pies and wraps = \$3.75 × 4 + \$2.90 × 2
= \$20.80

Cost of milk = \$23.65 − \$20.80

$$\begin{array}{r} 23.65 \\ -\ 20.80 \\ \hline 2.85 \end{array}$$

∴ the carton of milk cost \$2.85.

6. **\$3.70**

Cost of folder = \$1.20
Cost of ruler = \$0.60
Cost of pencil = \$0.30

Sam bought 2 rulers, 1 pencil and 4 folders.

Total cost = \$0.60 × 2 + \$0.30 + \$1.20 × 4
= \$6.30

Change = \$10.00 − \$6.30

$$\begin{array}{r} 10.00 \\ -\ \ 6.30 \\ \hline 3.70 \end{array}$$

∴ the change is \$3.70.

Revision Test 7
Level of difficulty—Average

(page 78)

1. **\$21.20**

Cost = \$3.20 + \$1.80 × 10
= \$3.20 + \$18
= \$21.20

∴ the cost is \$21.20.

2. **\$7.92**

Cost = 9.9 × 0.8
= 7.92

∴ the cost is \$7.92.

3. 4

$$\begin{aligned}\text{Number of laps} &= 2.4 \div 0.6\\ &= 24 \div 6\\ &= 4\end{aligned}$$

∴ she would complete 4 laps.

4. 3.9 kg

$$\begin{aligned}\text{Remainder} &= 10 - (3.2 + 2.9)\\ &= 10 - 6.1\\ &= 3.9\end{aligned}$$

∴ 3.9 kg of rice remains.

5. $4.80

$$\begin{aligned}\text{Change} &= 20 - (3.2 \times 3 + 2.8 \times 2)\\ &= 20 - (9.6 + 5.6)\\ &= 20 - 15.2\\ &= 4.8\end{aligned}$$

∴ her change is $4.80.

6. 97.85 kg

$$\begin{aligned}\text{Average} &= \frac{102.5 + 96.2 + 98.6 + 94.1}{4}\\ &= \frac{391.4}{4}\\ &= 97.85\end{aligned}$$

∴ the average is 97.85 kg.

Revision Test 8
Level of difficulty—Challenging (page 79)

1. 5.12

$$\begin{aligned}\text{Difference} &= 4.8 \div 0.6 - 4.8 \times 0.6\\ &= 48 \div 6 - 2.88\\ &= 8 - 2.88\\ &= 8.00 - 2.88\\ &= 5.12\end{aligned}$$

∴ the difference is 5.12.

2. 6.4 m

$$\begin{aligned}\text{Average of 5 distances} &= 7.2\\ \text{Total of 5 distances} &= 7.2 \times 5\\ &= 36\\ \text{Total of 4 distances} &= 6.8 + 7.5 + 8.1 + 7.2\\ &= 29.6\\ \text{5th distance} &= 36 - 29.6\\ &= 36.0 - 29.6\\ &= 6.4\end{aligned}$$

∴ the distance of the fifth competitor was 6.4 m.

3. 0.6 m

$$\begin{aligned}\text{Perimeter of each square} &= 4 \times 0.3\\ &= 1.2\\ \text{Wire used for 24 squares} &= 24 \times 1.2\\ &= 28.8\\ \text{Discarded length} &= 29.4 - 28.8\\ &= 0.6\end{aligned}$$

∴ Jack threw away 0.6 m of wire.

4. $80

$$\begin{aligned}\text{Cost per litre} &= 44.80 \div 28\\ &= 22.40 \div 14\\ &= 11.20 \div 7\\ &= 1.60\\ \text{Cost of Noel's petrol} &= 1.6 \times 50\\ &= 80.0\end{aligned}$$

∴ Noel spent $80.

5. 12.1 m

$$\begin{aligned}\text{Length of edging} &= 131.2 \div 3.2\\ &= 65.6 \div 1.6\\ &= 32.8 \div 0.8\\ &= 328 \div 8\\ &= 41\end{aligned}$$

∴ perimeter of garden bed is 41 m.

$$\begin{aligned}\text{length} + \text{width} &= 20.5\\ \text{length} + 8.4 &= 20.5\\ \text{length} &= 20.5 - 8.4\\ &= 12.1\end{aligned}$$

∴ length of garden bed is 12.1 m.

6. $390 000

$$\begin{aligned}\text{Amount Tracey receives} &= \$16\,000\\ \text{0.4 of Lacey's amount} &= \$16\,000\\ \text{0.1 of Lacey's amount} &= \$16\,000 \div 4\\ &= \$4000\\ \text{Whole of Lacey's amount} &= \$4000 \times 10\\ &= \$40\,000\\ \text{0.4 of Kacey's amount} &= \$40\,000\\ \text{0.1 of Kacey's amount} &= \$40\,000 \div 4\\ &= \$10\,000\\ \text{Whole of Kacey's amount} &= \$10\,000 \times 10\\ &= \$100\,000\\ \text{Total} &= \$16\,000 + \$40\,000 + \$100\,000\\ &= \$156\,000\\ \text{0.4 of total inheritance} &= \$156\,000\\ \text{0.1 of total inheritance} &= \$156\,000 \div 4\\ &= \$39\,000\\ \text{Whole of total inheritance} &= \$39\,000 \times 10\\ &= \$390\,000\end{aligned}$$

∴ the value of the inheritance was $390 000.

Key Skill 32 Percentages: Finding one quantity as percentage of another (pages 80–81)

1. 80%

$$\begin{aligned}\text{Percentage} &= \frac{16}{20} \times 100\%\\ &= \frac{\cancel{16}^{\,4}}{{}_1\cancel{5}\cancel{20}} \times \frac{\cancel{100}^{\,20}}{1}\%\\ &= 80\%\end{aligned}$$

∴ 80% of the chocolates had soft centres.

2. **75%**

$$\text{Percentage} = \frac{90}{120} \times 100\%$$
$$= \frac{^{3}\cancel{90}}{_{1}\cancel{4}\cancel{120}} \times \frac{\cancel{100}^{25}}{1}\%$$
$$= 75\%$$

∴ 75% of the students will compete.

3. **5%**

$$\text{Percentage} = \frac{3}{60} \times 100\%$$
$$= \frac{^{1}\cancel{3}}{_{1}\cancel{20}\cancel{60}} \times \frac{\cancel{100}^{5}}{1}\%$$
$$= 5\%$$

∴ Bruce's success rate was 5%.

4. **40%**

$$\text{Percentage} = \frac{240}{600} \times 100\%$$
$$= \frac{^{2}\cancel{240}}{_{1}\cancel{5}\cancel{600}} \times \frac{\cancel{100}^{20}}{1}\%$$
$$= 40\%$$

∴ 40% walk to school.

5. **75%**

$$\text{Percentage} = \frac{27}{36} \times 100\%$$
$$= \frac{^{3}\cancel{27}}{_{1}\cancel{4}\cancel{36}} \times \frac{\cancel{100}^{25}}{1}\%$$
$$= 75\%$$

∴ 75% of the cockatoos flew away.

6. **60%**

Total votes = 2229 + 12 000 + 5771

$$\begin{array}{r} 2\,229 \\ 12\,000 \\ +\ \ 5\,771 \\ \hline 20\,000 \end{array}$$

∴ total votes of 20 000

$$\text{Percentage} = \frac{12\,000}{20\,000} \times 100\%$$
$$= \frac{^{3}\cancel{12}}{_{1}\cancel{5}\cancel{20}} \times \frac{\cancel{100}^{20}}{1}\%$$
$$= 60\%$$

∴ 60% of the votes went to the winning candidate.

Key Skill  Percentages: Percentage change

(pages 82–83)

1. **20%**

$$\text{Increase} = \$24 - \$20$$
$$= \$4$$

$$\text{Percentage} = \frac{4}{20} \times 100\%$$
$$= \frac{^{1}\cancel{4}}{_{1}\cancel{5}\cancel{20}} \times \frac{\cancel{100}^{20}}{1}\%$$
$$= 20\%$$

∴ the calculator increased in price by 20%.

2. **10%**

$$\text{Decrease} = \$10 - \$9$$
$$= \$1$$

$$\text{Percentage} = \frac{1}{10} \times 100\%$$
$$= \frac{1}{_{1}\cancel{10}} \times \frac{\cancel{100}^{10}}{1}\%$$
$$= 10\%$$

∴ the magazine dropped in price by 10%.

3. **30%**

$$\text{Increase} = \$91 - \$70$$
$$= \$21$$

$$\text{Percentage} = \frac{21}{70} \times 100\%$$
$$= \frac{^{3}\cancel{21}}{_{1}\cancel{10}\cancel{70}} \times \frac{\cancel{100}^{10}}{1}\%$$
$$= 30\%$$

∴ Angelo's savings increased by 30%.

4. **25%**

$$\text{Decrease} = \$560 - \$420$$
$$= \$140$$

$$\text{Percentage} = \frac{140}{560} \times 100\%$$
$$= \frac{^{1}\cancel{2}\cancel{140}}{_{1}\cancel{4}\cancel{8}\cancel{560}} \times \frac{\cancel{100}^{25}}{1}\%$$
$$= 25\%$$

∴ the percentage discount was 25%.

5. **50%**

$$\text{Increase} = 36\,900 - 24\,600 = 12\,300$$

$$\text{Percentage} = \frac{12\,300}{24\,600} \times 100\% = \frac{^{1}\cancel{12\,300}}{_{1}\cancel{2}\cancel{24\,600}} \times \frac{\cancel{100}^{50}}{1}\% = 50\%$$

∴ there was a 50% increase in crowds.

6. **2%**

Increase = 210 000 − 175 000

$$\begin{array}{r} 210 \\ -\ 175 \\ \hline 35 \end{array}$$

∴ population increased by 35 000

$$\text{Percentage} = \frac{35\,000}{175\,000} \times 100\% = \frac{35\,\cancel{000}}{175\,\cancel{000}} \times \frac{100}{1}\% = \frac{^{1}\cancel{7}\cancel{35}}{_{5}\cancel{35}\cancel{175}} \times \frac{\cancel{100}^{20}}{1}\% = 20\%$$

$$\text{Average increase per year} = 20\% \div 10 = 2\%$$

∴ there was a 2% population increase each year.

Key Skill 34 Percentages: Fractions and percentages (pages 84–85)

1. $\mathbf{\frac{7}{10}}$

$$\text{Percentage male} = 100\% - 30\% = 70\%$$

$$\text{Fraction male} = \frac{70}{100} = \frac{7}{10}$$

∴ $\frac{7}{10}$ of passengers were male.

2. **20%**

$$\text{Fraction remaining} = 1 - \frac{4}{5} = \frac{1}{5}$$

$$\text{Percentage remaining} = \frac{1}{_{1}\cancel{5}} \times \frac{\cancel{100}^{20}}{1} = 20$$

∴ 20% of the tub remains.

3. $\mathbf{\frac{9}{20}}$

$$\text{Percentage girls} = 100\% - 55\% = 45\%$$

$$\text{Fraction girls} = \frac{45}{100} = \frac{9}{20}$$

∴ $\frac{9}{20}$ of babies were girls.

4. **72%**

$$\text{Number correct} = 25 - 7 = 18$$

$$\text{Fraction correct} = \frac{18}{25}$$

$$\text{Percentage correct} = \frac{18}{25} \times 100\% = \frac{18}{_{1}\cancel{25}} \times \frac{\cancel{100}^{4}}{1}\% = 72\%$$

∴ 72% of his answers were correct.

5. **40%**

$$\text{Fraction smokers} = \frac{3}{5}$$

$$\text{Fraction non-smokers} = 1 - \frac{3}{5} = \frac{2}{5}$$

$$\text{Percentage non-smokers} = \frac{2}{5} \times 100\% = \frac{2}{_{1}\cancel{5}} \times \frac{\cancel{100}^{20}}{1}\% = 40\%$$

∴ 40% of the pensioners never smoked.

6. $\mathbf{\frac{2}{5}}$

$$\text{Percentage swim} = 100\% - (25\% + 35\%) = 40\%$$

$$\text{Fraction swim} = \frac{40}{100} = \frac{2}{5}$$

∴ $\frac{2}{5}$ of those interviewed had swim leg as their favourite.

Key Skill  Percentages: Using percentages A

(pages 86–87)

1. **$2280**

   Amount = 30% of 7600
   = 0.3 × 7600
   = 2280.0
   = 2280

   ∴ $2280 is spent on the mortgage.

2. **$480**

   Amount = 20% of 2400
   = 0.2 × 2400
   = 480.0
   = 480

   ∴ $480 is the discount on the price of the cruise.

3. **18**

   Amount = 60% of 30
   = 0.6 × 30
   = 18.0

   ∴ 18 students have flown.

4. **4**

   Number = 5% of 80
   = 0.05 × 80
   = 4.00
   = 4

   ∴ Owen found 4 rotten apples.

5. **7.5 kg**

   Number = 15% of 50
   = 0.15 × 50
   = 7.50
   = 7.5

   ∴ Wendy's bones have a mass of 7.5 kg.

6. **36**

   Number correct = 90% of 40
   = 0.9 × 40
   = 36.0
   = 36

   ∴ Serena must get at least 36 questions correct.

Key Skill  Percentages: Using percentages B

(pages 88–89)

1. **78 cm**

   Increase = 30% of 60
   = 0.3 × 60
   = 18

   ∴ the plant increased in height by 18 cm.

   New height = 60 + 18
   = 78

   ∴ the plant is 78 cm high.

2. **192**

   Decrease = 20% of 240
   = 0.2 × 240
   = 48

   ∴ he reduces the number by 48.

   Remaining trees = 240 − 48
   = 192

   $$\begin{array}{r} 240 \\ -\ 48 \\ \hline 192 \end{array}$$

   ∴ 192 trees remain.

3. **$3.99**

   Increase = 5% of 3.80
   = 0.05 × 3.80
   = 0.19

   ∴ the shares increased by $0.19.

   New value = $3.80 + $0.19
   = $3.99

   ∴ the shares are now valued at $3.99 each.

4. **$384**

   Decrease = 20% of 480
   = 0.2 × 480
   = 96

   ∴ the price drops by $96.

   New price = $480 − $96
   = $384

   $$\begin{array}{r} 480 \\ -\ 96 \\ \hline 384 \end{array}$$

   ∴ the new price is $384.

5. **$2940**

   Increase = 40% of 2100
   = 0.4 × 2100
   = 840

   ∴ the price rose by $840

   New price = $2100 + 840
   = $2940

   ∴ the new price is $2940.

6. **$88**

   Increased price = $88 + $22
   = $110

   Decrease = 20% of 110
   = 0.2 × 110
   = 22

   ∴ the price drops by $22.

   New price = $110 − $22
   = $88

   $$\begin{array}{r} 110 \\ -\ 22 \\ \hline 88 \end{array}$$

   ∴ the new price is $88.

Key Skill 37 Percentages: Unitary method using percentages A

(pages 90–91)

1. **200**

8% of herd $= 16$
1% of herd $= 16 \div 8$
$= 2$
100% of herd $= 2 \times 100$
$= 200$
∴ Barry had 200 cows before the escape.

2. **\$400**

15% of price $= \$60$
1% of price $= \$60 \div 15$
$= \$4$
100% of price $= \$4 \times 100$
$= \$400$
∴ the dress originally cost \$400.

3. **\$1200**

5% of pay $= \$60$
1% of pay $= \$60 \div 5$
$= \$12$
100% of pay $= \$12 \times 100$
$= \$1200$
∴ Aaron was paid \$1200.

4. **\$600**

20% of last year's price $= \$120$
10% of last year's price $= \$120 \div 2$
$= \$60$
100% of last year's price $= \$60 \times 10$
$= \$600$
∴ gold cost \$600 last year.

5. **400**

12% of Paul's stickers $= 48$
1% of Paul's stickers $= 48 \div 12$
$= 4$
100% of Paul's stickers $= 4 \times 100$
$= 400$
∴ Paul has 400 stickers.

6. **\$1200**

New percentage $= 100\% - 40\%$
$= 60\%$
60% of original cost $= \$720$
10% of original cost $= \$720 \div 6$
$= \$120$
100% of original cost $= \$120 \times 10$
$= \$1200$
∴ with no discount it would have cost \$1200.

Key Skill 38 Percentages: Unitary method using percentages B

(pages 92–93)

1. **\$7000**

New price $= 100\% + 20\%$
$= 120\%$
120% of old price $= \$8400$
10% of old price $= \$8400 \div 12$
$= \$700$
100% of old price $= \$700 \times 10$
$= \$7000$
∴ the original price was \$7000.

2. **200**

Percentage remaining $= 100\% - 30\%$
$= 70\%$
70% of balloons $= 140$
10% of balloons $= 140 \div 7$
$= 20$
100% of balloons $= 20 \times 10$
$= 200$
∴ there were originally 200 balloons.

3. **600**

Percentage remaining $= 100\% - 40\%$
$= 60\%$
60% of original total $= 360$
10% of original total $= 360 \div 6$
$= 60$
100% of original total $= 60 \times 10$
$= 600$
∴ Idris originally had 600 apples.

4. **5 tonnes**

Percentage remaining $= 100\% - 20\%$
$= 80\%$
Using 4 tonnes $= 4000$ kg,
80% of load $= 4000$
10% of load $= 4000 \div 8$
$= 500$
100% of load $= 500 \times 10$
$= 5000$
∴ 5000 kg, or 5 tonnes was originally delivered.

5. **\$40**

New cost $= 100\% + 10\%$
$= 110\%$
110% of old cost $= \$44$
10% of old cost $= \$44 \div 11$
$= \$4$
100% of old cost $= \$4 \times 10$
$= \$40$
∴ the cost was \$40 before the tip.

6. **50 L**

$$\begin{aligned} \text{Difference} &= 80\% - 20\% \\ &= 60\% \\ 60\% \text{ of total capacity} &= 30 \\ 10\% \text{ of total capacity} &= 30 \div 6 \\ &= 5 \\ 100\% \text{ of total capacity} &= 5 \times 10 \\ &= 50 \end{aligned}$$

∴ total capacity is 50 litres.

Revision Test 9
Level of difficulty—Average (page 94)

1. **27**

$$\begin{aligned} \text{Percentage present} &= 100\% - 10\% \\ &= 90\% \\ \text{Number present} &= 90\% \text{ of } 30 \\ &= 0.9 \times 30 \\ &= 27.0 \end{aligned}$$

∴ 27 students are present.

2. **$27.50**

$$\begin{aligned} \text{Total income} &= \$900 + \$800 + \$1050 = \$2750 \\ \text{Donation} &= 1\% \text{ of } 2750 \\ &= 0.01 \times 2750 \\ &= \$27.50 \end{aligned}$$

∴ $27.50 is donated.

3. **$24**

$$\begin{aligned} \text{Toni's rent} &= \text{one-third of } \$720 \\ &= \$720 \div 3 \\ &= \$240 \\ \text{Yvette's rent} &= 30\% \text{ of } \$720 \\ &= 0.3 \times \$720 \\ &= \$216.0 \\ \text{Difference} &= \$240 - \$216 \\ &= \$24 \end{aligned}$$

∴ Toni pays $24 more than Yvette.

4. **$1248**

$$\begin{aligned} \text{Increase} &= 4\% \text{ of } \$1200 \\ &= 0.04 \times \$1200 \\ &= \$48 \\ \text{New pay} &= \$1200 + \$48 \\ &= \$1248 \end{aligned}$$

∴ Tyler's new pay is $1248.

5. **25%**

$$\begin{aligned} \text{Remaining money} &= \$240 - (\$120 + \$60) \\ &= \$240 - \$180 \\ &= \$60 \\ \text{Percentage} &= \frac{60}{240} \times 100 \\ &= \frac{1}{{}_1\cancel{4}} \times \frac{\cancel{100}^{25}}{1} \\ &= 25 \end{aligned}$$

∴ Meg has 25% remaining.

6. **$624 000**

$$\begin{aligned} \text{Increase} &= 30\% \text{ of } \$480\,000 \\ &= 0.3 \times \$480\,000 \\ &= \$144\,000.0 \\ \text{Present value} &= \$480\,000 + \$144\,000 \\ &= \$624\,000 \end{aligned}$$

∴ the present value is $624 000.

Revision Test 10
Level of difficulty—Challenging (page 95)

1. **$30**

As 90 + 70 = 160, then

Martha sells $\frac{9}{16}$ and Mary sells $\frac{7}{16}$.

$$\begin{aligned} \text{Martha's sales} &= \frac{9}{16} \text{ of } 240 \\ &= \frac{9}{{}_{1}\cancel{2}\cancel{16}} \times \frac{\cancel{240}^{\cancel{30}\,15}}{1} \\ &= 135 \\ \text{Mary's sales} &= 240 - 135 \\ &= 105 \\ \text{Difference} &= 135 - 105 \\ &= 30 \end{aligned}$$

∴ Martha made $30 more than Mary.

2. **54**

$$\begin{aligned} \text{Scored in first 30 games} &= 30\% \text{ of } 30 \\ &= 0.3 \times 30 \\ &= 9 \\ \text{Scored in next 40 games} &= 40\% \text{ of } 40 \\ &= 0.4 \times 40 \\ &= 16 \\ \text{Scored in final 10 games} &= 10\% \text{ of } 10 \\ &= 1 \\ \text{Games not scored} &= 80 - (9 + 16 + 1) \\ &= 80 - 26 \\ &= 54 \end{aligned}$$

∴ Barry did not score in 54 games.

3. **$168**

$$\begin{aligned} \text{Increase in watch} &= \$480 - \$400 \\ &= \$80 \\ \text{Percentage increase} &= \frac{80}{400} \times 100 \\ &= \frac{1}{{}_1\cancel{5}} \times \frac{\cancel{100}^{20}}{1} \\ &= 20\% \\ \text{Bangle price increase} &= 20\% \text{ of } 140 \\ &= 0.2 \times 140 \\ &= 28.0 \\ \text{Bangle new price} &= \$140 + \$28 \\ &= \$168 \end{aligned}$$

∴ the bangle's new price is $168.

4. **18 L**

Amount of concentrate $= 60\%$ of 12
$= 0.6 \times 12$
$= 7.2$

$\therefore$ the mixture contains 7.2 L of concentrate.

40% of new mixture $= 7.2$
10% of new mixture $= 7.2 \div 4$
$= 1.8$
100% of new mixture $= 1.8 \times 10$
$= 18$

$\therefore$ Trent needs an 18-litre container.

5. **306**

40% of total girls $= 220$
10% of total girls $= 220 \div 4$
$= 55$
100% of total girls $= 550$
Total number of boys $= 1060 - 550$
$= 510$
Number of boy reps $=$ 60% of 510
$= 0.6 \times 510$
$= 306.0$

$\therefore$ 306 boys represented the school.

6. **50%**

160% of cost price $= 320$
10% of cost price $= 320 \div 16$
$= 20$
100% of cost price $= 20 \times 10$
$= 200$
Employee cost $=$ 80% of 200
$= 0.8 \times 200$
$= 160$

$\therefore$ the employee cost is \$160

Amount of discount $= \$320 - \$160$
$= \$160$

Discount percentage $= \frac{160}{320} \times 100$
$= \frac{1}{2} \times 100$
$= 50$

$\therefore$ the employee discount is 50%.

ALGEBRA

Key Skill  39 Using pronumerals

(page 96–97)

1. $q(p + r)$

Kilograms of prawns $= p + r$
Mass of prawns $= q \times (p + r)$

$\therefore$ Leon bought $q(p + r)$ prawns, or $(pq + qr)$ prawns.

2. **$(100q - (pt + bw))$ cents**

There are $100q$ cents in \$$q$.

Total cost $= t \times p + w \times b$
$= pt + bw$
Change $= 100q - (pt + bw)$

$\therefore$ change is $(100q - (pt + bw))$ cents.

3. $k(p - q)$

Number correct $= p - q$
Marks $= (p - q) \times$ k
$= k(p - q)$

$\therefore$ Glen scored $k(p - q)$ marks.

4. **$(3b + d)$ years old**

Alvin's age $= b$
Kevin's age $= 3 \times b$
$= 3b$
Lennie $= 3b + d$

$\therefore$ Lennie is $(3b + d)$ years old.

5. **$(w - t + y)$ kg**

(Let's pretend there are numbers instead of pronumerals: Robert is 50, Jack is 50 − 4 and Max is 50 − 4 + 2).

Jack's mass $= w - t$
Max's mass $= w - t + y$

$\therefore$ Max has a mass of $(w - t + y)$ kg.

6. $(\frac{2x}{3} - y)$

Ming kept $\frac{2}{3}$ of x. This is written as $\frac{2x}{3}$ oranges. Of these, she used y oranges.

$\therefore$ She has $(\frac{2x}{3} - y)$ oranges remaining.

Key Skill 40 Equations A

(pages 98–99)

1. **13**

Let Meg's age be x.
This means Ken's age is $x + 5$.

$x + x + 5 = 31$
$2x + 5 = 31$
$2x + 5 - 5 = 31 - 5$
$\frac{2x}{2} = \frac{26}{2}$
$x = 13$

$\therefore$ Meg is 13 years old.

2. **50 cents**
Let the cost of an apple be x cents.
The cost of an orange is $(x + 20)$ cents.
$$2 \times x + x + 20 = 170$$
$$2x + x + 20 = 170$$
$$3x + 20 - 20 = 170 - 20$$
$$\frac{3x}{3} = \frac{150}{3}$$
$$x = 50$$
∴ an apple costs 50 cents.

3. **2 kg**
You can write 3.2 kg as 3200 grams.
Let the smaller container have x grams.
Let the larger container have $(x + 800)$ grams.
$$x + x + 800 = 3200$$
$$2x + 800 = 3200$$
$$2x + 800 - 800 = 3200 - 800$$
$$\frac{2x}{2} = \frac{2400}{2}$$
$$x = 1200$$
The smaller container has 1200 g. As 1200 + 800 is 2000, the larger container has 2000 g.
∴ the larger container has 2 kg of rice.

4. **25 minutes**
Let the time he swims be x minutes.
Let the time he runs be $2x$ minutes.
$$2x + x = 75$$
$$3x = 75$$
$$\frac{3x}{3} = \frac{75}{3}$$
$$x = 25$$
∴ Armir swims for 25 minutes each morning.

5. **21 and 28**
Let the number in smaller class be x.
Let the number in larger class be $x + 7$.
$$x + x + 7 = 49$$
$$2x + 7 = 49$$
$$2x + 7 - 7 = 49 - 7$$
$$2x = 42$$
$$\frac{2x}{2} = \frac{42}{2}$$
$$x = 21$$
∴ the classes have 21 and 28 students.

6. **pencil 40c, pen 70c**
Let the cost of a pencil be x cents.
Let the cost of a pen be $(x + 30)$ cents.
$$5 \times x + 5 \times 30 + 2 \times x = 430$$
$$5x + 150 + 2x = 430$$
$$7x + 150 = 430$$
$$7x + 150 - 150 = 430 - 150$$
$$7x = 280$$
$$\frac{7x}{7} = \frac{280}{7}$$
$$x = 40$$
∴ a pencil costs 40c and pen 70c.

Key Skill  Equations B

(pages 100–101)

1. **24 cm and 12 cm**
Let the breadth be x.
Let the length be $2x$.
$$2 \times (2x + x) = 72$$
$$2 \times 3x = 72$$
$$6x = 72$$
$$\frac{6x}{6} = \frac{72}{6}$$
$$x = 12$$
∴ the dimensions are 24 cm and 12 cm.

2. **80°, 70° and 30°**
Let the largest angle be x.
Let the middle-sized angle be $(x - 10)°$.
Let the smallest angle be $(x - 50)°$.
$$x + x - 10 + x - 50 = 180$$
$$3x - 60 = 180$$
$$3x - 60 + 60 = 180 + 60$$
$$3x = 240$$
$$\frac{3x}{3} = \frac{240}{3}$$
$$x = 80$$
∴ the angles are 80°, 70° and 30°.

3. **7 cm, 17 cm and 21 cm**
Let the smallest side be x.
Let the second side be $3x$.
Let the third side be $x + 10$.
$$x + 3x + x + 10 = 45$$
$$5x + 10 = 45$$
$$5x + 10 - 10 = 45 - 10$$
$$5x = 35$$
$$\frac{5x}{5} = \frac{35}{5}$$
$$x = 7$$
∴ the sides are 7 cm, 17 cm and 21 cm.

4. **30°, 60° and 90°**
Let the smallest angle be $x°$.
Let the middle-sized angle be $2x°$.
Let the largest angle be $3x°$.
$$x + 2x + 3x = 180$$
$$6x = 180$$
$$\frac{6x}{6} = \frac{180}{6}$$
$$x = 30$$
∴ the angles are 30°, 60° and 90°.

5. **6 cm**
Let the side of the square be x.
Let the length of the rectangle be $x + 4$.
Let the width of the rectangle be $x - 4$.
$2 \times (x + 4 + x - 4) = 24$
$2 \times 2x = 24$
$4x = 24$
$\frac{4x}{4} = \frac{24}{4}$
$x = 6$
$\therefore$ the sides of the square are 6 cm.

6. **80 cm²**
Let the width be x.
Let the length be $2x - 6$.
$2 \times (2x - 6 + x) = 36$
$2 \times (3x - 6) = 36$
$6x - 12 = 36$
$6x - 12 + 12 = 36 + 12$
$6x = 48$
$\frac{6x}{6} = \frac{48}{6}$
$x = 8$
Length $= 2x - 6$
$= 2 \times 8 - 6$
$= 16 - 6$
$= 10$
$\therefore$ the sides are 10 cm and 8 cm
$\therefore$ the area is 80 cm^2.

Revision Test 11
Level of difficulty—Average (page 102)

1. **$(pt + qv)$ cents**
Cost $= p \times t + q \times v$
$= pt + qv$
$\therefore$ the cost is $(pt + qv)$ cents.

2. **$(d + k + 20)$ years old**
Aretha's age $= d$
Artha's mother's age $= d + 20$
Aretha's aunt $= d + 20 + k$
$\therefore$ Aretha's aunt is $(d + k + 20)$ years old.

3. **\$5**
Let the cost of a hamburger be x.
The price of a juice is $x - 3$.
$2 \times x + x - 3 = 12$
$2x + x - 3 = 12$
$3x - 3 = 12$
$3x - 3 + 3 = 12 + 3$
$3x = 15$
$\frac{3x}{3} = \frac{15}{3}$
$x = 5$
$\therefore$ a hamburger costs \$5.

4. **\$120**
Let the oldest sister have x.
The middle sister has $x - 40$
The youngest sister has $x - 10$.
$x + x - 40 + x - 10 = 340$
$3x - 50 = 340$
$3x - 50 + 50 = 340 + 50$
$3x = 390$
$\frac{3x}{3} = \frac{390}{3}$
$x = 130$
The youngest sister has $130 - 10 = 120$.
$\therefore$ she has \$120.

5. **80°, 40° and 60°**
Let the largest angle be x.
The smallest angle is $x - 40$.
The other angle is $x - 40 + 20 = x - 20$.
$x + x - 40 + x - 20 = 180$
$3x - 60 = 180$
$3x - 60 + 60 = 180 + 60$
$3x = 240$
$\frac{3x}{3} = \frac{240}{3}$
$x = 80$
$\therefore$ angles are 80°, 40° and 60°.

6. **\$50**
Let Keith's amount be x.
Rita put in $x - 20$.
Jane put in $x + 10$.
$x + x - 20 + x + 10 = 200$
$3x - 10 = 200$
$3x - 10 + 10 = 200 + 10$
$3x = 210$
$\frac{3x}{3} = \frac{210}{3}$
$x = 70$
Rita put in $70 - 20 = 50$.
$\therefore$ Rita contributed \$50.

Revision Test 12
Level of difficulty—Challenging
(page 103)

1. **$[(a + 2d) \times (b + 2d) - ab]$ m²**

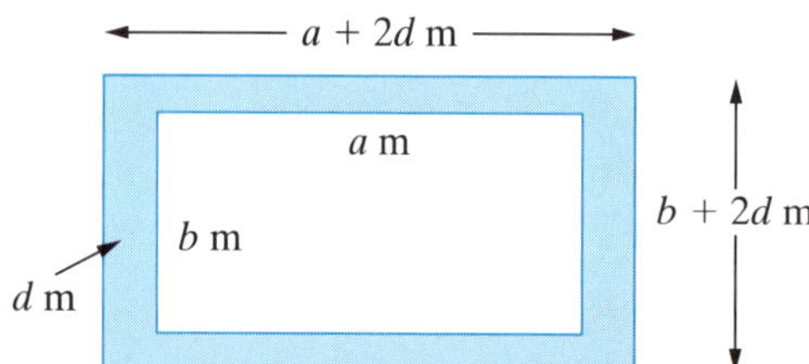

2. **33**
Let Harriet's age be x.
Melissa is $2x$.
Lionel is $x + 6$.

$$\begin{aligned} x + 2x + x + 6 &= 42 \\ 4x + 6 &= 42 \\ 4x + 6 - 6 &= 42 - 6 \\ 4x &= 36 \\ \frac{4x}{4} &= \frac{36}{4} \\ x &= 9 \end{aligned}$$

Harriet is 9, Melissa is 18 and Lionel is 15.
When Melissa is 36 Lionel will be $36 - 3 = 33$.
$\therefore$ Lionel will be 33 years old.

3. **8**
Edward got 72%, or 36 out of 50.
Let the number of questions correct be x.
Edward got $10 - x$ questions wrong.

$$\begin{aligned} 5 \times x - 2 \times (10 - x) &= 36 \\ 5x - 20 + 2x &= 36 \\ 7x - 20 + 20 &= 36 + 20 \\ 7x &= 56 \\ \frac{7x}{7} &= \frac{56}{7} \\ x &= 8 \end{aligned}$$

$\therefore$ Edward answered 8 questions correctly.

4. **\$4.30**
Let the cost of Brand B be x.
Brand A cost is $2x$.
Brand C cost is $2x - 270$.

$$\begin{aligned} x + 2x + 2x - 270 &= 1480 \\ 5x - 270 &= 1480 \\ 5x - 270 + 270 &= 1480 + 270 \\ 5x &= 1750 \\ \frac{5x}{5} &= \frac{1750}{5} \\ x &= 350 \end{aligned}$$

Brand B costs \$3.50.
Brand C costs $2 \times \$3.50 - \$2.70 = \$4.30$
$\therefore$ Brand C costs \$4.30.

5. **\$124.80**
Perimeter of 144 m.
$\therefore$ length + width = 72 m
Let width be x.
Length is $2 \times x + 12 = 2x + 12$.

$$\begin{aligned} 2x + 12 + x &= 72 \\ 3x + 12 &= 72 \\ 3x + 12 - 12 &= 72 - 12 \\ 3x &= 60 \\ \frac{3x}{3} &= \frac{60}{3} \\ x &= 20 \end{aligned}$$

$\therefore$ the length is $2 \times 20 + 12 = 52$
$\therefore$ length 52 cm and width 20 cm.

Depth = 60 cm = 0.6 m

$$\begin{aligned} \text{Volume} &= 52 \times 20 \times 0.6 \\ &= 1040 \times 0.6 \\ &= 624.0 \end{aligned}$$

$\therefore$ volume of mulch is 624 m^3

$$\begin{aligned} \text{Cost} &= 624 \times \$0.20 \\ &= \$124.80 \end{aligned}$$

$\therefore$ cost of mulch is \$124.80.

6. **56**
Five years ago, let Annie's age be x.
Also, 5 years ago, Garry was $3x$.
Now, Annie is $(x + 5)$ and Garry is $(3x + 5)$.
In 12 years: Annie will be $(x + 17)$ and Garry will be $(3x + 17)$.

$$\begin{aligned} 3x + 17 &= 2 \times (x + 17) \\ 3x + 17 &= 2x + 34 \\ x &= 17 \end{aligned}$$

$\therefore$ Garry is $3x + 5$ or $3 \times 17 + 5 = 56$.

MEASUREMENT

Key Skill 42 Area A (pages 104–105)

1. **36 m²**

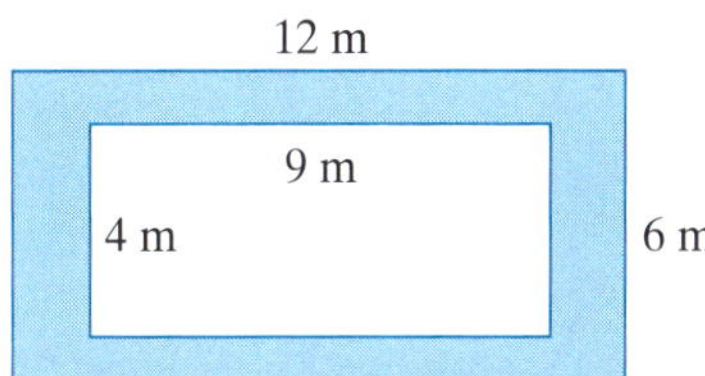

$$\begin{aligned} \text{Area} &= 12 \times 6 - 9 \times 4 \\ &= 72 - 36 \\ &= 36 \end{aligned}$$

$\therefore$ the area is 36 m^2.

2. **480 cm²**

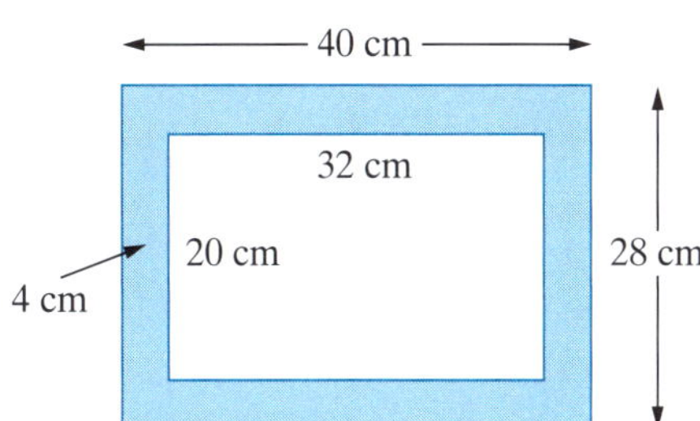

Area = $40 \times 28 - 32 \times 20$

$$\begin{array}{r} 28 \\ \times\ 40 \\ \hline 1120 \end{array} \qquad \begin{array}{r} 32 \\ \times\ 20 \\ \hline 640 \end{array} \qquad \begin{array}{r} 1120 \\ -\ 640 \\ \hline 480 \end{array}$$

$\therefore$ the area is 480 cm^2.

3. **19 900 cm²**

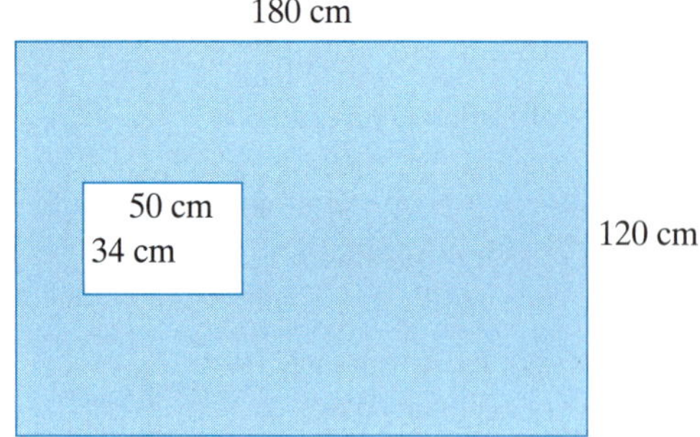

Area = 180 × 120 − 50 × 34
= 21 600 − 1700
= 19 900
∴ the area is 19 900 cm².

4. **32 m²**

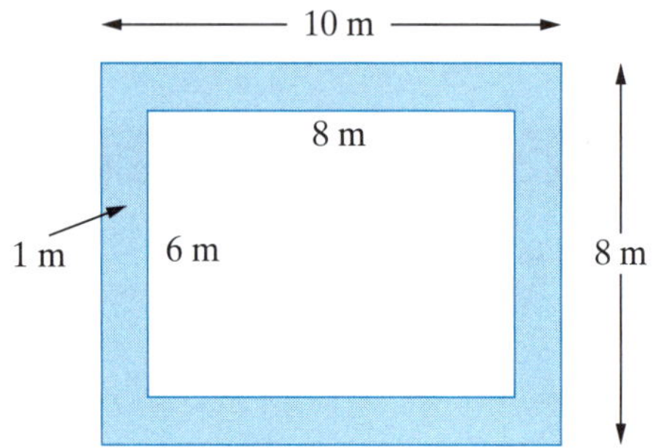

Area = 10 × 8 − 8 × 6
= 80 − 48
= 32
∴ The area is 32 m².

5. **11 340 cm²**

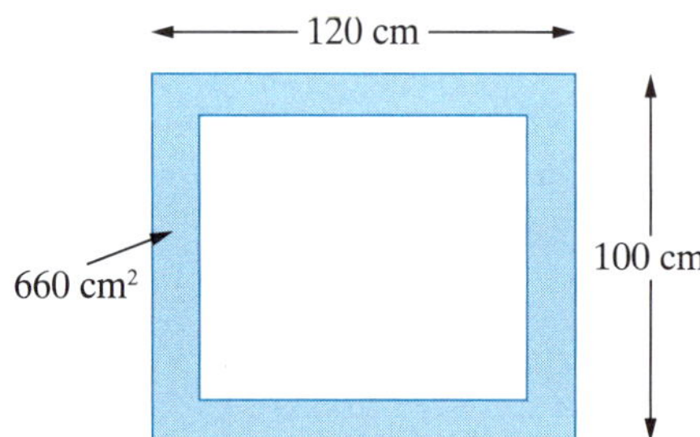

Area = 120 × 100 − 660
= 12 000 − 660
= 11 340
∴ the area is 11 340 cm².

6. **60 m²**

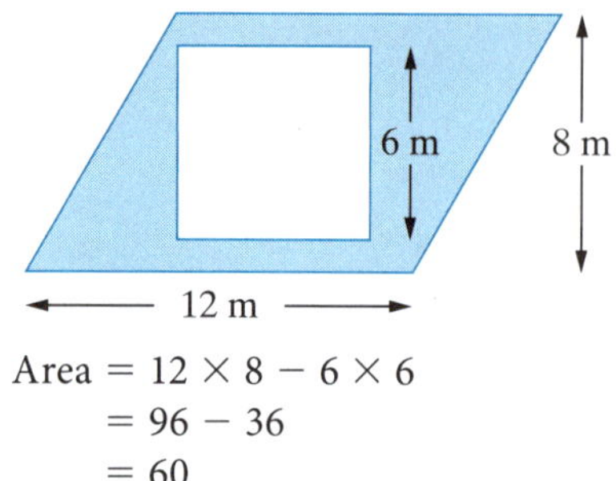

Area = 12 × 8 − 6 × 6
= 96 − 36
= 60
∴ the grassed area is 60 m².

Key Skill  Area B (pages 106–107)

1. **$972**
Area = 9 × 9
= 81
Cost = 81 × 12
= 972
∴ the cost is $972.

2. **$1920**
Area = 2 × 6 × 4
= 48
Cost = 48 × 40
= 1920
∴ the cost is $1920.

3. **$2520**
Area = 450 × 280
= 900 × 140
= 126 000
∴ the area is 126 000 m²
Cost = 126 000 × 0.02
= 2520
∴ the cost is $2520.

4. **$595**
Area = $\frac{1}{2}$ × 3.4 × 5
= 1.7 × 5
= 8.5
Cost = 8.5 × 70
= 595.0
∴ the cost is $595.

5. **1 965 600**
Area = 210 × 120
= 25 200
Number = 25 200 × 78

```
    25 200
 ×      78
   201 600
 1 764 000
 1 965 600
```

∴ there are 1 965 600 plants.

6. **\$208**
As 2 m = 200 cm and 1.6 m = 160 cm, then
$200 \div 20 = 10$ and $160 \div 20 = 8$.
Number of squares $= 10 \times 8$
$= 80$
Cost $= \$2.60 \times 80$
$= \$208.00$
$\therefore$ the cost of the wool was \$208.

Key Skill  Area C (pages 108–109)

1. **\$840**
Area $= 350 \times 200$
$= 70\,000$
Cost $= 70\,000 \div 1000 \times 12$
$= 70 \times 12$
$= 840$
$\therefore$ the cost is \$840.

2. **6 L**
Area $= 12 \times 8$
$= 96$
Litres $= 96 \div 8 \times 0.5$
$= 12 \times 0.5$
$= 6$
$\therefore$ 6 litres of oil is required.

3. **96**
Area $= 30 \times 20$
$= 600$
Mats $= 600 \div 25 \times 4$
$= 2400 \div 100 \times 4$
$= 24 \times 4$
$= 96$
$\therefore$ Mr Graham needs 96 mats.

4. **280 mL**
Area $= \frac{1}{2} \times 40 \times 28$
$= 560$
Quantity $= 560 \div 10 \times 5$
$= 56 \times 5$
$= 280$
$\therefore$ Graeme needs 280 mL.

5. **\$720**
Use 40 cm = 0.4 m.
To find the number of tiles:
$3.6 \div 0.4 = 36 \div 4 = 9$ and
$4 \div 0.4 = 40 \div 4 = 10$.
Number of tiles $= 9 \times 10$
$= 90$
Cost $= 90 \times \$8$
$= \$720$
$\therefore$ the tiles will cost \$720.

6. **\$56**
Area $= 5.5 \times 5$
$= 27.5$
Litres $= 27.5 \div 12$ $\quad 12\overline{)27.5} = 2.292$
$\therefore$ Lee needs about 2.3 litres.
If he buys three 1-L tins it will cost \$63. It is better to buy the 4-L tin for \$56.

Key Skill  Volume A (pages 110–111)

1. **28.8 m^3**
Volume $= 8 \times 6 \times 0.6$
$= 48 \times 0.6$
$= 28.8$
$\therefore$ Dylan needs 28.8 m^3.

2. **74.4 cm^3**
Area $= 6.2 \times 12$
$= 74.4$
$\therefore$ the volume is 74.4 cm^3

3. **968 cm^3**
Volume $= 22 \times 11 \times 4$
$= 242 \times 4$
$= 968$
$\therefore$ the paver has a volume of 968 cm^3.

4. **2400 cm^3**
Change 2 metres to 200 cm.
Volume $= 200 \times 12$
$= 2400$
$\therefore$ the volume of the bar is 2400 cm^3.

5. **16 m^3**
Volume $= 4 \times 2 \times 2$
$= 16$
$\therefore$ the volume of the container is 16 m^3.

6. **150 mm**
Volume of original $= 600 \times 100 \times 100$
$= 6\,000\,000$
$\therefore$ the volume of original is 6 000 000 mm^3
Volume of new = length $\times$ width $\times$ height
$6\,000\,000 = 200 \times 200 \times$ height
$6\,000\,000 = 40\,000 \times$ height
height $= 6\,000\,000 \div 40\,000$
$= 600 \div 4$
$= 150$
$\therefore$ the height of the new ingot would be 150 mm.

Key Skill 46 Volume B (pages 112–113)

1. **7.5 kg**
   Volume = $5 \times 5 \times 5$
   = 125
   ∴ the volume of metal is 125 cm^3.
   Mass = 125×60
   = 7500
   ∴ the mass of the cube is 7500 grams, or 7.5 kg.

2. **400 cm^3**
   Volume = $30 \times 20 \times 10$
   = 6000
   ∴ the volume of cake is 6000 cm^3.
   Volume each person = $6000 \div 15$
   = 400
   ∴ each person receives 400 cm^3.

3. **64 cm^3**
   Volume = $16 \times 10 \times 8$
   = 1280
   ∴ the volume of clay is 1280 cm^3.
   Volume each student = $1280 \div 20$
   = 64
   ∴ each student receives 64 cm^3.

4. **3.2 m^3**
   Volume = $4 \times 2 \times 1.6$
   = 1.6×8
   = 12.8
   ∴ the volume of soil is 12.8 m^3.
   Volume each person = $12.8 \div 4$
   = 3.2
   ∴ each person receives 3.2 m^3.

5. **3.2 kg**
   Volume = $20 \times 10 \times 8$
   = 1600
   ∴ the volume of seeds is 1600 cm^3.
   Mass = 1600×20
   = 3200
   ∴ the mass of seeds is 3200 grams, or 3.2 kg.

6. **600 g**
   Volume of first box = $25 \times 20 \times 6$
   = 500×6
   = 3000
   ∴ volume of cereal in first box is 3000 cm^3.
   Volume each gram = $3000 \div 750$
   = $300 \div 75$
   = 4
   ∴ the cereal has a mass of 1 gram per 4 cm^3.
   Volume of second box = $20 \times 15 \times 8$
   = 300×8
   = 2400
   ∴ volume of cereal in second box is 2400 cm^3.
   Mass of second box = $2400 \div 4$
   = 600
   ∴ the second box contains 600 grams.

Revision Test 13
Level of difficulty—Average (page 114)

1. **48 m^2**

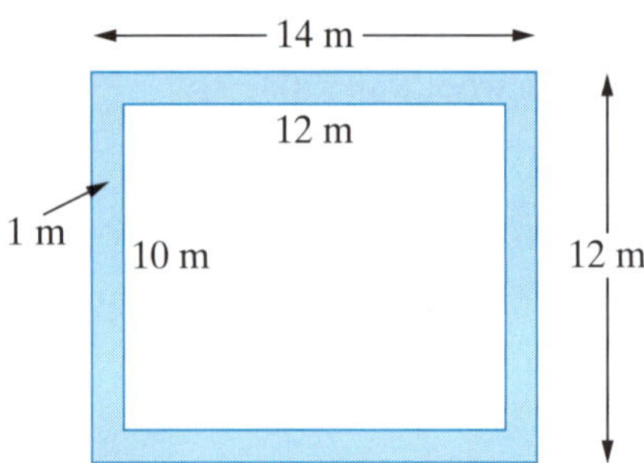

   Area = $14 \times 12 - 12 \times 10$
   = $168 - 120$
   = 48
   ∴ the area of the grass border is 48 m^2.

2. **$50.40**
   Area = $\frac{1}{2} \times 4.2 \times 3$
   = 2.1×3
   = 6.3
   Cost = 6.3×8
   = 50.4
   ∴ the cost is $50.40.

3. **5.4 m^2**

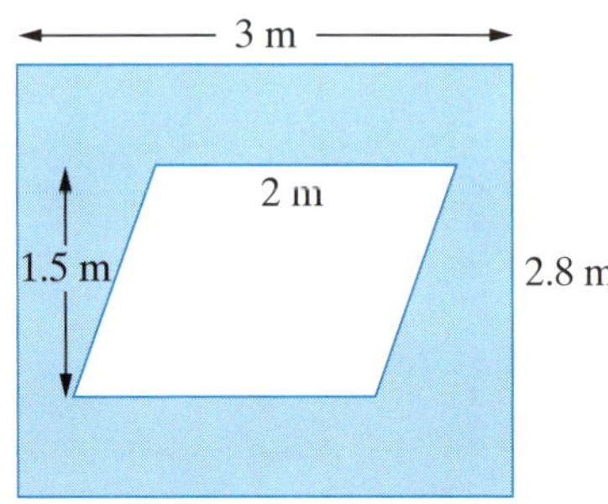

   Area = $3 \times 2.8 - 2 \times 1.5$
   = $8.4 - 3$
   = 5.4
   ∴ 5.4 m^2 of the wall is not gold.

4. **90 m^3**
   Volume = $30 \times 10 \times 0.3$
   = 300×0.3
   = 90.0
   ∴ the volume is 90 m^3.

5. **4.5 L**
   Area of wall = 3×3
   = 9
   Total area of 2 coats = 18
   Amount of paint = $18 \div 4$
   = 4.5
   ∴ Arissa needs 4.5 litres.

6. 27 cm^3

Area of square face = 9
Length of square's side = 3
Volume = 3 × 3 × 3
= 27

∴ the cube's volume is 27 cm^3.

Revision Test 14
Level of difficulty—Challenging

(page 115)

1. 5 cm

Area = base × perpendicular height
32 = 6.4 × perpendicular height
perpendicular height = 32 ÷ 6.4
= 320 ÷ 64
= 5

∴ perpendicular height is 5 cm.

2. 125 cm^3

No. of edges on cube = 12
Length of each edge = 60 ÷ 12
= 5

∴ edge length is 5 cm
Volume = 5 × 5 × 5
= 125

∴ volume is 125 cm^3.

3. 4 cm

Dimensions are length, width, height.
We can rewrite as 3 × width, width, 2 × width
∴ volume = 3 width × width × 2 width
48 = 6 × (width)^3
(width)^3= 8
∴ width = 2

This means length = 6 cm and height = 4 cm.
∴ height is 4 cm.

4. 300 m

Area = 1440 ÷ 4 × 100
= 360 × 100
= 36 000

Area = length × width
36 000 = length × 120
Length = 36 000 ÷ 120
= 3600 ÷ 12
= 300

∴ length is 300 m.

5. 40

1 m = 100 cm.
50 × **2** = 100
25 × **4** = 100
20 × **5** = 100
Number = 2 × 4 × 5
= 40

∴ Penny can fit 40 boxes in the carton.

6. 37.5%

Volume of soil in trailer = 1.6 × 2 × 0.6
= 3.2 × 0.6
= 1.92

Volume of soil in garden = 2 × 2 × 0.3
= 4 × 0.3
= 1.2

Amount remaining = 1.92 − 1.2
= 0.72

$$\text{Percentage remaining} = \frac{0.72}{1.92} \times 100 = \frac{72}{192} \times 100 = \frac{6}{16} \times 100 = \frac{3}{8} \times 100 = 37.5$$

∴ 37.5% remains.

STATISTICS

Key Skill 47 Averages A

(pages 116–117)

1. 4.1 km

$$\text{Average} = \frac{5 + 4.2 + 3.7 + 4.6 + 3}{5} = \frac{20.5}{5} = 4.1$$

∴ the average distance was 4.1 km.

2. $70

Average of 4 weeks = $60
Total of 4 weeks = $240
Average of next 2 weeks = $90
Total of next 2 weeks = $180
Total of 6 weeks = $240 + $180
= $420
Average of 6 weeks = $420 ÷ 6
= $70

∴ Miriam was paid an average of $70.

3. 11 mm

$$\text{Average} = \frac{16 + 8 + 10 + 5 + 16}{5} = \frac{55}{5} = 11$$

∴ the average daily rainfall was 11 mm.

4. **3**

Average of 3 months = 2
Total of 3 months = 6
Average of next 3 months = 4
Total of next 3 months = 12
Total of 6 months = 6 + 12
= 18
Average of 6 weeks = 18 ÷ 6
= 3

∴ Leo read an average of 3 books per month.

5. **14**

$$\text{Average} = \frac{18 + 12 + 15 + 0 + 16 + 23}{6}$$
$$= \frac{84}{6}$$
$$= 14$$

∴ the average was 14 points per game.

6. **$632**

Average price of 4 steers = $650
Total price of 4 steers = $650 × 4
= $2600
Average price of 6 heifers = $620
Total price of 6 heifers = $620 × 6
= $3720
Total price of 10 cattle = $2600 + $3720
= $6320
Average price of 10 cattle = $632

∴ the average price was $632.

Key Skill  48 Averages B

(pages 118–119)

1. **21**

Average of 4 tests = 16
Sum of 4 tests = 16 × 4
= 64
Average of 5 tests = 17
Sum of 5 tests = 17 × 5
= 85
5th test result = 85 − 64
= 21

∴ Dave needs 21 in his next test.

2. **70 kg**

Average of 6 people = 60
Sum of 6 people = 60 × 6
= 360
Average of 5 people = 58
Sum of 5 people = 58 × 5
= 290
Karen's mass = 360 − 290
= 70

∴ Karen's mass is 70 kg.

3. **48**

Average of 5 innings = 24
Sum of 5 innings = 24 × 5
= 120
Average of 6 innings = 24 + 4
= 28
Sum of 6 innings = 28 × 6
= 168
6th innings score = 168 − 120
= 48

∴ he scored 48.

4. **4 cm**

Average of 3 straws = 8
Sum of 3 straws = 8 × 3
= 24
Average of 2 straws = 10
Sum of 2 straws = 10 × 2
= 20
Length of 3rd straw = 24 − 20
= 4

∴ the third straw is 4 cm long.

5. **181 cm**

Average of 5 players = 182
Sum of 5 players = 182 × 5
= 910
Sum of 4 players = 910 − 186
= 724
Average of 4 players = 724 ÷ 4
= 181

∴ average height of 4 players is 181 cm.

6. **7**

Average of 6 children = 10
Sum of 6 children = 10 × 6
= 60
Sum of 5 children = 60 − 5
= 55
Average of 4 boys = 12
Sum of 4 boys = 12 × 4
= 48
Age of other girl = 55 − 48
= 7

∴ the other girl is 7 years old.

Key Skill 49 Diagrams, tables and guess-and-check A

(pages 120–121)

1. **5**

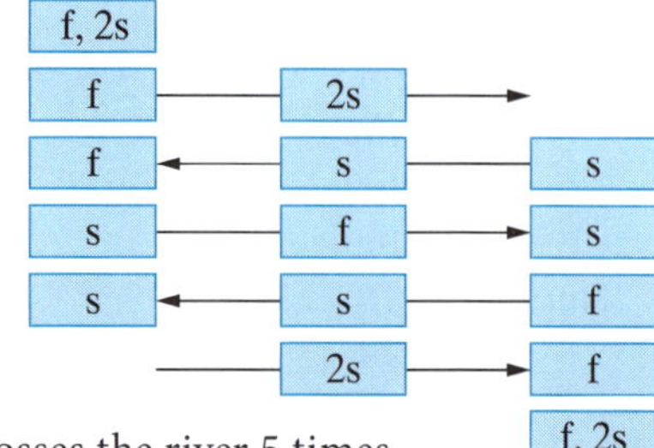

∴ the boat crosses the river 5 times.

2. **16**

Guess	3-legged	4-legged	Legs
1	13	13	39 + 52 = 91
2	12	14	36 + 56 = 92
3	10	16	30 + 64 = 94

∴ he made 16 four-legged stools (and 10 three-legged).

3. **11**

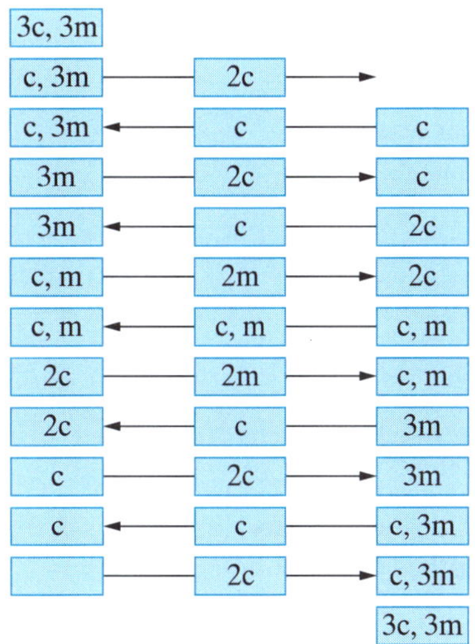

∴ the boat crosses the river 11 times.

4. **2 spiders and 8 beetles**

Guess	Spider (8)	Beetle (6)	Legs
1	5	5	40 + 30 = 70
2	3	7	24 + 42 = 66
3	2	8	16 + 48 = 64

∴ she has 2 spiders and 8 beetles.

5. **24**

Guess	5c coins	10c coins	Total value
1	20	20	\$1 + \$2 = \$3
2	15	25	\$0.75 + \$2.50 = \$3.25
3	16	24	\$0.80 + \$2.40 = \$3.20

∴ Jackson has 24 ten-cent coins.

6. **5**

Pig (\$20)	Chicken (\$5)	Sheep (\$30)	Total cost
40	30	20	\$800 + \$150 + \$600 = \$1550
10	80	10	\$200 + \$400 + \$300 = \$900
20	60	20	\$400 + \$300 + \$600 = \$1300
15	70	15	\$300 + \$350 + \$450 = \$1100
20	70	10	\$400 + \$350 + \$300 = \$1050
25	70	5	\$500 + \$350 + \$150 = \$1000

∴ Dave will buy 5 sheep (also, 25 pigs and 70 chickens).

Key Skill 50: Diagrams, tables and guess-and-check B (pages 122–123)

1. **15**

Guessed	3 ×	÷ 3	Sum	Sum + 9
3	9	1	10	19
6	18	2	20	29
15	45	5	50	59

∴ the number is 15.

2. **108**

Red	Green	Chocolate	Liquorice
			12
108	48	24	12

∴ there are 108 red frogs in the shop.

3. **36**

Guess	Janeel	Kai	Penny	Owen	Total
1	24	12	6	4	46
2	40	20	10	8	78
3	36	18	9	7	70

∴ Janeel is 36 years old.

4. **\$100**

Guess	Anne	Bennie	Sam	Total
1	\$60	\$40	\$90	\$190
2	\$80	\$60	\$110	\$250
3	\$70	\$50	\$100	\$220

∴ Sam receives \$100.

5. **$2100**

Guess	Bill 1	Bill 2	Bill 3	Diff. 1 & 3
1	$150	$50	$500	$350
2	$1500	$500	$5000	$3500
3	$300	$100	$1000	$700
4	$900	$300	$3000	$2100
5	$450	$150	$1500	$1050

Total = $450 + $150 + $1500
= $2100
∴ the total of the three bills is $2100.

6. **$84**

Guessed	Top: One-third − $5	Skirt: Half + $15	Change
$60	$15	$45	$0
$66	$17	$48	$1
$90	$25	$60	$5
$84	$23	$57	$4

∴ she was given $84 for her birthday.

Revision Test 15
Level of difficulty—Average (page 124)

1. **2 km**

$$\text{Average} = \frac{3 + 2.5 + 1.8 + 2.1 + 0.6}{5} = \frac{10}{5} = 2$$

∴ the average distance was 2 km.

2. **99%**

Average on 4 tests = 84
Total on 4 tests = 84 × 4
= 336
Average on 5 tests = 87
Total on 5 tests = 87 × 5
= 435
Fifth test result = 435 − 336
= 99
∴ John must score 99% on his next test.

3. **159.5 cm**

Average of 3 students = 152
Total of 3 students = 152 × 3
= 456
Average of other 5 students = 164
Total of other 5 students = 164 × 5
= 820
Total of 8 students = 456 + 820
= 1276
Average of 8 students = 1276 ÷ 8
= 159.5
∴ the average height was 159.5 cm.

4. **8**

Let the number be x

$\therefore 3 \times x + 18 - 12 = 30$
$3x + 6 = 30$
$3x + 6 - 6 = 30 - 6$
$3x = 24$
$\frac{3x}{3} = \frac{24}{3}$
$x = 8$

∴ the number was 8.

5. **12 chickens**

Guess	Chicken (2)	Pig (4)	Legs
1	10	10	20 + 40 = 60
2	12	8	24 + 32 = 56

∴ the farmer has 12 chickens and 8 pigs.

6. **12 hours**

Let the hours worked by Horatio be x.
∴ Harold worked for $2x$ hours,
Henry worked $(2x - 6)$ hours.

$x + 2x + 2x - 6 = 54$
$5x - 6 = 54$
$5x - 6 + 6 = 54 + 6$
$5x = 60$
$\frac{5x}{5} = \frac{60}{5}$
$x = 12$

∴ Horatio worked for 12 hours.

Revision Test 16
Level of difficulty—Challenging (page 125)

1. **1**

Average of 8 scores = 12
Sum of 8 scores = 12 × 8
= 96
Average of 10 scores = 12 − 2
= 10
Sum of 10 scores = 10 × 10
= 100
Total of 2 new scores = 100 − 96
= 4
If one of the new scores is 3, then the other new score is 1.
∴ the other new score is 1.

2. Penny

Steps	Murray	Penny
1	3	19
2	6	21
3	9	23
4	12	25
5	15	27
6	18	29
7	21	31
8	24	33
9	27	35
10	30	37
11	33	39
12	36	Finished
13	39	
14	Finished	

∴ Penny reaches the top first.

3. 50 kg

Ken + Jen = 90
Ken + Ben = 60
∴ Jen is 30 kg heavier than Ben
Jen + Ben = 50
∴ Jen must be 40 kg and Ben is 10 kg.
∴ Ken has a mass of 50 kg.

4. Day 4

Start: −9
Day 1: −9 + 4 = −5
Night 1: −5 − 2 = −7
Day 2: −7 + 4 = −3
Night 2: −3 − 2 = −5
Day 3: −5 + 4 = −1
Night 3: −1 − 2 = −3
Day 4: −3 + 4 = 1
The frog is out!
∴ the frog gets out on Day 4.

5. 15 minutes

Angela (A), Brock (B), Colin (C), Dorrie (D):

Left bank	Bridge	Right bank	Time
ABCD			
CD	AB →		2
CD	← B	A	2
B	CD →		8
B	← A	CD	1
	AB →	CD	2
		ABCD	

∴ the fastest time is 15 minutes

6. 5

9:00 am: 1. 7:00 am train just arrives
9:30 am: 2. 8:00 am train passes
10:00 am: 3. 9:00am train passes
10:30 am: 4. 10:00am train passes
11:00 am: 5. 11:00 am train just leaves
∴ there are 5 trains that will be met.

NOTES